Supporting Air and Space Expeditionary Forces

Lesso
Operation
Iraqi
Freedom

Kristin F. Lynch, John G. Drew,
Robert S. Tripp, C. Robert Roll, Jr.

Prepared for the United States Air Force

Approved for public release; distribution unlimited

RAND PROJECT AIR FORCE

The research reported here was sponsored by the United States Air Force under contract F49642-01-C-0003. Further information may be obtained from the Strategic Planning Division, Directorate of Plans, Hq USAF.

Library of Congress Cataloging-in-Publication Data

Lessons from Operation Iraqi Freedom / Kristin F. Lynch ... [et al.].
    p. cm.—(Supporting air and space expeditionary forces)
    Includes bibliographical references.
    "MG-193."
    ISBN 0-8330-3677-7 (pbk.)
    1. United States. Air Force—Supplies and stores. 2. Iraq War, 2003—Case studies.
    3. Airlift, Military—United States. 4. Deployment (Strategy) I. Lynch, Kristin.
    II. Series.

    UG1123.L47 2005
    358.4'1621'0973—dc22

                                                                    2004024811

The RAND Corporation is a nonprofit research organization providing objective analysis and effective solutions that address the challenges facing the public and private sectors around the world. RAND's publications do not necessarily reflect the opinions of its research clients and sponsors.

RAND® is a registered trademark.

Cover design by Barbara Angell Caslon

Published 2005 by the RAND Corporation
1776 Main Street, P.O. Box 2138, Santa Monica, CA 90407-2138
1200 South Hayes Street, Arlington, VA 22202-5050
201 North Craig Street, Suite 202, Pittsburgh, PA 15213-1516
RAND URL: http://www.rand.org/
To order RAND documents or to obtain additional information, contact
Distribution Services: Telephone: (310) 451-7002;
Fax: (310) 451-6915; Email: order@rand.org

Preface

Since 1997, the RAND Corporation has studied options for configuring a future agile combat support (ACS) system that would enable the achievement of Air and Space Expeditionary Force (AEF) goals. Operation Iraqi Freedom (OIF), the war in Iraq, offered opportunities to examine the implementation of new ACS concepts in a wartime environment. In 2000, Project AIR FORCE helped evaluate combat support lessons from Joint Task Force Noble Anvil (JTF NA),[1] the air war over Serbia, part of Operation Allied Force (OAF). In 2002, Project AIR FORCE evaluated combat support lessons from Operation Enduring Freedom (OEF), in Afghanistan. Some of the concepts and lessons learned from JTF NA and OEF were implemented in supporting OIF.

Supporting Air and Space Expeditionary Forces: Lessons from Operation Iraqi Freedom presents an analysis of combat support experiences associated with Operation Iraqi Freedom and compares these experiences with those associated with Operation Allied Force and Operation Enduring Freedom.[2] The analysis presented an opportunity to compare findings and implications from JTF NA, OEF, and OIF. Specifically, the objectives of the analysis were to describe how combat support was accomplished in OIF, examine

---

[1] Joint Task Force Noble Anvil was the code name for U.S. forces involved in Operation Allied Force. This report concentrates on Air Force operations conducted by Joint Task Force Noble Anvil.

[2] This report does not address medical support issues.

how ACS concepts were implemented in OIF, and compare JTF NA, OEF, and OIF experiences to determine similarities and applicability of lessons across experiences and to determine whether some experiences are unique to particular scenarios.

This analysis concentrates on U.S. Air Force operations in support of OAF (JTF NA), the first 100 days of OEF, and the year of planning as well as the first month of OIF. It was conducted to address the above objectives, and thus the report focuses on experiences from OIF and what those experiences imply for combat support system design to ensure that AEF goals can be achieved. It does not address medical support issues.

The Commander, Air Combat Command (ACC/CC) sponsored this research, which was conducted in the Resource Management Program of RAND Project AIR FORCE, in coordination with the Air Force Deputy Chief of Staff for Installations and Logistics (AF/IL). The research for this report was completed in August 2003.

This report should be of interest to logisticians, operators, and mobility planners throughout the Department of Defense, especially those in the Air Force.

This study is one of a series of RAND reports that address ACS issues in implementing the AEF. Other publications in the series include the following:

- *Supporting Expeditionary Aerospace Forces: An Integrated Strategic Agile Combat Support Planning Framework*, Robert S. Tripp, Lionel A. Galway, Paul S. Killingsworth, Eric Peltz, Timothy L. Ramey, and John G. Drew (MR-1056-AF). This report describes an integrated combat support planning framework that may be used to evaluate support options on a continuing basis, particularly as technology, force structure, and threats change.
- *Supporting Expeditionary Aerospace Forces: New Agile Combat Support Postures*, Lionel Galway, Robert S. Tripp, Timothy L. Ramey, and John G. Drew (MR-1075-AF). This report describes how alternative resourcing of forward operating locations (FOLs) can support employment timelines for future

AEF operations. It finds that rapid employment for combat requires some prepositioning of resources at FOLs.

- *Supporting Expeditionary Aerospace Forces: An Analysis of F-15 Avionics Options*, Eric Peltz, H. L. Shulman, Robert S. Tripp, Timothy L. Ramey, Randy King, and John G. Drew (MR-1174-AF). This report examines alternatives for meeting F-15 avionics maintenance requirements across a range of likely scenarios. The authors evaluate investments for new F-15 avionics intermediate shop test equipment against several support options, including deploying maintenance capabilities with units, performing maintenance at forward support locations (FSLs), or performing all maintenance at the home station for deploying units.

- *Supporting Expeditionary Aerospace Forces: A Concept for Evolving to the Agile Combat Support/Mobility System of the Future*, Robert S. Tripp, Lionel A. Galway, Timothy L. Ramey, Mahyar A. Amouzegar, and Eric Peltz (MR-1179-AF). This report describes a vision for the ACS system of the future based on individual commodity study results.

- *Supporting Expeditionary Aerospace Forces: Expanded Analysis of LANTIRN Options*, Amatzia Feinberg, H. L. Shulman, L. W. Miller, and Robert S. Tripp (MR-1225-AF). This report examines alternatives for meeting Low Altitude Navigation and Targeting Infrared for Night (LANTIRN) support requirements for AEF operations. The authors evaluate investments for new LANTIRN test equipment against several support options including deploying maintenance capabilities with units, performing maintenance at FSLs, or performing all maintenance at continental United States support hubs for deploying units.

- *Supporting Expeditionary Aerospace Forces: Alternatives for Jet Engine Intermediate Maintenance*, Mahyar A. Amouzegar, Lionel A. Galway, and Amanda Geller (MR-1431-AF). This report evaluates the manner in which Jet Engine Intermediate Maintenance (JEIM) shops can best be configured to facilitate overseas deployments. The authors examine a number of JEIM support options, which are distinguished primarily by the degree to which JEIM support is centralized or decentralized.

- *Reconfiguring Footprint to Speed Expeditionary Aerospace Forces Deployment*, Lionel Galway, Mahyar A. Amouzegar, R. J. Hillestad, and Don Snyder (MR-1625-AF). This study develops an analytic framework—footprint configuration—to assist in evaluating the feasibility of reducing the size of equipment or time-phasing the deployment of support and relocating some equipment to places other than forward operating locations. It also attempts to define *footprint* and to establish a way to monitor its reduction.

- *Supporting Expeditionary Aerospace Forces: An Operational Architecture for Combat Support Execution Planning and Control*, James A. Leftwich, Robert S. Tripp, Amanda Geller, Patrick H. Mills, Tom LaTourrette, Charles Robert Roll, Cauley Von Hoffman, and David Johansen (MR-1536-AF). This report outlines a framework for evaluating options for combat support execution planning and control. The analysis describes the combat support command and control operational architecture as it is now and as it should be in the future. It also describes the changes that must take place to achieve that future state.

- *Supporting Air and Space Expeditionary Forces: Lessons from Operation Enduring Freedom*, Robert S. Tripp, Kristin F. Lynch, John G. Drew, and Edward W. Chan (MR-1819-AF). This report describes the expeditionary ACS experiences during the war in Afghanistan and compares them with experiences in Joint Task Force Noble Anvil to determine similarities and unique practices. It indicates how well the ACS framework performed during these contingency operations. From this analysis, the ACS framework may be updated to better support the AEF concept.

RAND Project AIR FORCE

RAND Project AIR FORCE (PAF), a division of the RAND Corporation, is the U.S. Air Force's federally funded research and development center for studies and analyses. PAF provides the Air Force with independent analyses of policy alternatives affecting the development, employment, combat readiness, and support of current and future aerospace forces. Research is performed in four programs: Aerospace Force Development; Manpower, Personnel, and Training; Resource Management; and Strategy and Doctrine.

Additional information about PAF is available on our website at http://www.rand.org/paf.

Contents

Figures

Tables

Summary

The Air Force developed the Air and Space Expeditionary Force (AEF) concept—substituting speed of deployment and employment for presence—to allow it to respond quickly to any national security issue with a tailored, sustainable force. Since 1997, RAND Project AIR FORCE and the Air Force Logistics Management Agency have studied and refined a framework for an agile combat support (ACS) system to support the AEF concept (Galway et al., 2000; Tripp et al., 1999).

Agile Combat Support System Background

As described in Tripp et al. (2000), the AEF operational goals are to

- foster an expeditionary mind-set
- rapidly configure support needed to achieve the desired operational effects
- quickly deploy both large and small tailored force packages with the capability to deliver substantial firepower anywhere in the world
- immediately employ such forces upon arrival
- smoothly shift from deployment to operational sustainment
- meet the demands of small-scale contingencies and peacekeeping commitments while maintaining readiness for potential contingencies outlined in defense guidance.

Key elements of an ACS system to enable these AEF operational goals include the following (Tripp et al., 1999):

- An expeditionary—forward-thinking—mind-set, which would be instilled in combat support personnel
- A combat support execution planning and control (CSC2) system to assess, organize, and direct combat support[3] activities, meet operational requirements, and be responsive to rapidly changing circumstances. The CSC2 capability would help combat support personnel
  —estimate combat support resource requirements and process performance needed to achieve the desired operational effects for the specific scenario.
  —configure supply chains for deployment and sustainment, including the military and commercial transportation needed to meet deployment and sustainment needs.
  —establish control parameters for the performance of various combat support processes required to meet specific operational needs.
  —track actual combat support performance against control parameters.
  —signal when a process is outside accepted control parameters so that plans can be developed to get the process back within control limits.
- A quickly configured and responsive distribution network to connect forward operating locations (FOLs), forward support locations (FSLs), and continental United States (CONUS) support locations (CSLs)
- A network of FOLs resourced to support varying deployment/ employment timelines

---

[3] In this report, the term *combat support* is defined as anything other than the actual flying operation. Combat support consists of civil engineering, communications, security forces, maintenance, service, munitions, etc. Not all aspects of combat support are addressed in this report because the scope was too broad.

- A network of FSLs configured outside CONUS to provide storage capabilities for heavy war reserve materiel (WRM), such as munitions and tents, and selected maintenance capabilities, such as centralized intermediate repair facilities (CIRFs) that service jet engines of units deployed to FOLs. FSLs could be collocated with transportation hubs.
- A network of CSLs, including Air Force depots, CIRFs, and contractor support facilities. As with FSLs, a variety of different activities may be set up at major Air Force bases, convenient civilian transportation hubs, or Air Force or other defense repair depots.

In 2000, Project AIR FORCE helped evaluate combat support lessons from Joint Task Force Noble Anvil (JTF NA),[4] in Serbia. In 2002, it evaluated combat support lessons from Operation Enduring Freedom (OEF), in Afghanistan. Some of the concepts and lessons learned from JTF NA and OEF were implemented in supporting Operation Iraqi Freedom (OIF).

Concentrating only on Air Force operations, this analysis provided the opportunity to compare findings and implications from JTF NA, OEF, and OIF. Specifically, the objectives of the analysis were to indicate how combat support performed in OIF, examine how ACS concepts were implemented in OIF, and compare JTF NA, OEF, and OIF experiences to determine similarities and applicability of lessons across experiences and to determine whether some experiences are unique to particular scenarios. This report does not address medical support issues.

JTF NA, OEF, and OIF provide three important opportunities to study how AEF ACS concepts were implemented during contingency operations and how they have been refined with each contingency experience to better support AEF goals. All three contingency operations provide important opportunities to study

---

[4] The U.S. portion of Operation Allied Force was code named Joint Task Force Noble Anvil. This report concentrates on Air Force operations conducted by JTF NA.

how AEF ACS concepts were implemented during contingency operations. In this report, we address six areas: CSC2 structure, FOL development, the use of FSLs and CSLs, the transportation system, the use of current technology, and resourcing to meet current operational requirements. Understanding these experiences could be of value for combat support and operational personnel who may be called upon to support future contingency operations. The Commander, Air Combat Command (ACC/CC), sponsored this research in coordination with the Air Force Deputy Chief of Staff for Installations and Logistics (AF/IL).

Combat Support Execution Planning and Control

Findings (see pp. 19–38)

Lessons learned from JTF NA indicated problems in combat support execution planning and control. As a result, AF/IL asked RAND Project AIR FORCE to develop a CSC2 future, or "TO-BE," operational architecture.[5] That work began in 2000 and was concluded just as operations began in Afghanistan. Although the TO-BE operational architecture was not used during OEF, OEF provided an opportunity to improve its design.

Many of the issues identified in JTF NA and OEF did not emerge during OIF, because doctrine was undergoing change before OIF and roles and responsibilities were being defined or redefined. Likewise, because many standing organizations were still in place from OEF and because the command structure for those organizations was well defined, individuals and organizations were better prepared to meet their responsibilities than they were during Joint Task Force Noble Anvil. The leaders in OIF had learned from the recently completed OEF. Many of the same leaders held their same positions for OIF. Organizations built on an ad hoc basis for OEF were refined for OIF. Early in OIF planning, roles and

---

[5] See the Appendix for a list of CSC2 TO-BE nodes and their responsibilities.

responsibilities were clearly defined and articulated. Given the long planning time, a solid plan was developed. Moreover, U.S. Air Forces, Central Command (CENTAF), was the supported command, with the rest of the Air Force supporting CENTAF. Although numerous other operations were ongoing, combat support personnel focused on providing the combatant commander with the essential elements he needed to succeed in OIF.

To address essential elements of the Logistics Sustainability Analysis (LSA) that was used to help build the combat support plan for OIF, the Air Force Materiel Command (AFMC) hosted a Warfighter Support Conference, which AF/IL facilitated. The LSA identified actions that needed to be taken to support the operational plan. Many major commands (MAJCOMs) and functions were represented and helped to finalize the plan. Once the efforts were all focused, it became easier for all involved to reach an agreement on the type and format of information that needed to flow between echelons. Standardized reports were defined and web-based updates were used.

The ability to adjust a plan during execution can become the most important requirement during any operation. The expeditionary mind-set of OIF leaders aided the success of the operation. Although the Air Force has had equipment and personnel deployed in the area of responsibility (AOR) since the end of Operation Desert Storm, problems occurred during OIF once the combatant commander changed the sequence of the forces called forward. In such a changing environment, the success of CENTAF during OIF was due to motivated, highly trained, ingenious individuals working around problems within the system.

Implications (see pp. 38–41)

The Air Force should ensure that the lessons from this operation—both the good and the bad—are passed on to future leaders, perhaps through doctrinal changes. Doctrine should institutionalize success from past operations. An expeditionary Air Force will be required in the future. Training and equipping leaders to deal with expeditionary operations will continue to be a challenge.

Much should be learned from the way in which roles and responsibilities were defined during OIF. Organizations established in recent operations should become standing organizations, used regularly.

Combat support planning needs to be integrated in the operational campaign planning process. The effects of alternative combat support strategies, tactics, and configurations need to be known to operations personnel when a plan is selected. In addition, once a jointly developed operation and combat support plan has been determined to be feasible—that is, capable of achieving the desired operational effects—a closed-loop[6] feedback and control system needs to track actual combat support process performance against planned values. When the system exceeds control parameter limits, the CSC2 system needs to signal combat support personnel that corrective action is needed. The CSC2 operational architecture outlines how this planning and control could occur across the echelons of support and throughout the phases of operational campaigns (Leftwich et al., 2002).

The combat support execution planning and control operational architecture also specifies CSC2 nodes and associated responsibilities that are consistent with those that were developed and used during OIF. The CSC2 architecture specifies the broad responsibilities of a commander, Air Force forces (COMAFFOR) A-4 (Logistics Directorate of the Air Component Staff, or A-Staff[7]) Forward and Rear, the Contingency Support Center, and Inventory Control Points. Many of the COMAFFOR A-4 functions, as well as those of the other nodes, can be performed by standing organizations.

Command and control reachback support needs to be defined for all A-staff functions. Should reachback be separated by functional responsibility, or should all A-staff functions be collocated in standing rear organizations that can serve more than one COMAFFOR?

---

[6] A *closed-loop process* takes the output and uses it as an input for the next iteration of the process.

[7] The term "A-staff" refers to an Air Force staff that is organized using the joint staff designation (J-1, J-2, etc.).

Collocated reachback is the option presented in the CSC2 TO-BE operational architecture. For instance, if another numbered Air Force were to conduct a sizable operation at the same time that 9th Air Force was engaged in OIF, Air Combat Command (ACC) could conduct reachback for the new operation from a new section of the same Operations Support Center.

Implementing the CSC2 operational architecture concepts requires changes in doctrine, education and training, organization, and systems. AF/IL has initiated doctrinal changes to begin implementation of the CSC2 processes and standing organizations—a step in the right direction. But much more is needed. Education and training programs are needed to teach these concepts. The expeditionary mind-set should be incorporated in all doctrine, policy, education, and training so that leaders, both current and future, are prepared for expeditionary operations.

In addition, decision support systems are needed to carry control information to combat support personnel so that significant deviations from planned performance can be corrected for before operational effects are felt. The use of information systems has improved, but additional capability is needed, including automated system interfaces through which better access to control information could be provided. The wider use of automated tools would enhance beddown[8] assessments. Better links are needed between operational requirements and AFMC process performance and resource levels.

JOPES, the systems that support development of Time Phased Force and Deployment Data (TPFDD) in the deliberate planning cycle, should also be able to support crisis action execution. One possible solution may be for the Air Force to offer a single unit type code (UTC) to the combatant commander, as the Army does, and then to internally tailor that UTC as required. Another option may be Force Modules, whereby a prearranged force would be able to provide a given capability. Yet another option is a new system whereby certain inputs yield indexed outputs. For example, the

---

[8] *Beddown* refers to the basing locations of personnel and/or aircraft during operations.

number of fighters or bombers a commander wants to deploy are the inputs and the output is a list of UTCs; or the input could be a capability (the ability to provide close air support for a given size army force for X days, and deep strike capability to destroy Y number of targets and all associated expeditionary combat support to bed down and sustain forces for Z days) so the tool would not just add UTCs but would look for redundancy and define the requirements. CSC2 is a vital component of agile combat support concepts needed to meet air and space expeditionary force operational goals.

Forward Operating Locations and Site Preparation

Findings (see pp. 43–58)

Build-up timelines for forward operating locations varied in OIF and depended heavily on the preparation activities. Those FOLs that were partially developed or at which the Air Force had experience in previous deployments facilitated rapid force deployments. Many of the FOLs developed in support of OEF were also used in OIF. The speed with which FOLs could become operational depended on country clearances, access to the real estate, quality and timeliness of the site survey, the amount of development needed to bed down forces, and the amount of contract support available, among other things. For Air Force planners to have had detailed knowledge of the AOR before OIF greatly enhanced their ability to open bases.

During OIF, the decision to move Air Force forces forward into Iraq created additional challenges. Basic necessities, such as fuel, water, rations, housing, and rapid runway repair, all had to be brought into the country.

Preparation of FOLs was slowed by host-nation support. Even when host nations agreed to allow forces to use their facilities, they often asked that their support not become public knowledge. In some countries, such as Turkey, the support that the planners had assumed they would receive did not materialize.

Civil engineering played a large role in getting OIF forward operating locations ready for deploying forces. Civil engineers as well as resources were stressed in buildup efforts for ongoing deployments and for new construction efforts. There was also a large buildup in communications in the AOR. Finally, contractor support facilitated FOL development. Selection and development of FOLs play an important role in meeting the air and space expeditionary force goal.

Implications (see pp. 59–60)

In all three recent military operations, large amounts of time were expended in gaining country and specific-FOL access. Even when FOL sites were known and anticipated, time was required to develop these sites. Engagement policies and programs to familiarize Air Force planners with facilities in countries that may be sites for future operations could potentially reduce country access time. Such programs as Partnership for Peace, in which knowledge of and improvements to FOLs can be gained through exercises and deployments, could be valuable and should be encouraged. Knowledge gained through this and other programs that enhance military-to-military contact could help speed deployments to important areas around the world.

Where possible, a select number of future FOLs in likely sites should be surveyed for capabilities. Goals could be established in each area of responsibility for surveying potential sites for future Air Force use, and funds could be set aside for carrying out such surveys. In some cases, sites in potential conflict areas could be prepared in advance for rapid deployment.

Training some Air Force combat support officers similarly to Army Foreign Area Officers could produce some country and area specialists.[9] Foreign Area Specialists could augment embassies in the early stages of conflict, when military staffs at embassies are often overwhelmed. They could facilitate rapid country clearances, access, and host-nation support agreements. In addition, military staffs at embassies should be augmented during wartime. During OIF, the

---

[9] The Air Force does have a career-broadening duty similar to the Foreign Area Officer. There is also discussion about developing a more robust Professional Military Strategist program in the Air Force for language and cultural specialists.

augmentation to many of the embassies was at the General Officer level, alleviating many potential roadblocks.

Once a contingency begins, leveraging contractor capabilities to assist civil engineers in developing FOLs is another method of decreasing FOL preparation time. The Air Force Contract Augmentation Program and other contractor capabilities, such as WRM maintenance contractors at forward support locations, can be capitalized on to aid civil engineers in rapidly building up and then sustaining forward operating locations, as demonstrated in OIF. Although it may be desirable to have Air Force civil engineers complete the initial beddown planning and construction, capabilities to augment scarce Air Force personnel skills could be developed through these programs. Databases of contractor capabilities, similar to FOL site surveys, should be developed in areas where potential conflicts may be likely.

Forward Support Location/CONUS Support Location Preparation for Meeting Uncertain FOL Requirements

Findings (see pp. 61–69)
Combat support resources, including fuel, munitions, spare parts, and rations, dominated sustainment movements.

As in JTF NA and OEF, moving assets from forward support locations to the forward operating locations satisfied most FOL combat support requirements. If speed of delivery of materiel is a requirement in future operations and the throughput issues are not resolved, the potential throughput constraints identified at some forward support locations during both OIF and OEF could slow deployment of large forces.

CONUS support locations were used effectively during OIF. During OEF, attention was given to creating better links between CSLs and the warfighters. AFMC has a Logistics Support Office and created a High Impact Target list to enhance responsiveness to the warfighter. AFMC expanded the Logistics Support Office and the High Impact Target list during OIF.

Centralized intermediate repair facilities were used successfully during OEF and again during OIF. They satisfied intermediate maintenance requirements for a number of reparable items for deployed fighter units, reducing the forward deployed footprint. Goals were established linking war fighter needs to the performance of the CIRF maintenance process and the theater distribution system.

Implications (see pp. 69–70)

JTF NA, OEF, and OIF demonstrate that future conflicts are likely to occur far from CONUS. A global network of FSLs with prepositioned WRM is necessary to meet AEF goals. The use of austere FOLs and an immature theater infrastructure in both OEF and OIF has illustrated the need for a portfolio of FSLs. The current AEF force structure of light, lean, and lethal response forces is highly dependent on forward support locations.

When developing a portfolio of FSLs to support numerous different operational challenges, many options should be provided and available for use in future contingencies. Trade-offs between improving existing FSLs, which may enhance throughput and storage capacity, and developing new FSLs need to be examined.

When considering whether to develop new FSLs or improve existing facilities, attention should be given to joint requirements. All services depend upon prepositioned materiel to meet contingency requirements. The management of joint facilities to meet multiple-service requirements may reduce operating costs. Information needs to be shared among services as well as with U.S. allies. If such arrangements are pursued, the throughput required for all participants needs to be considered explicitly.

Since the centralized intermediate repair facility Concept of Operations has been successful in the past two operations, CIRFs will likely be more widely used in future operations. As CIRFs are used in more operations, their requirements for reliable transportation should be included in the planning process. The trade-off of reducing deployment airlift in the early stages of a conflict is the availability of reliable sustainment transportation beginning on Day 1 of the operation. Without assured airlift, CIRFs will struggle to meet AEF operational goals. More work is required to ensure that the combatant commanders understand and support the risk that the Air Force is taking when agreeing to maintain aircraft using CIRF.

Reliable Transportation to Meet FOL Needs

Findings (see pp. 71–89)

Fuel dominated movement requirements. Assets such as FOL support, munitions, and rations also accounted for a significant portion of movements. Although spares accounted for only a small portion of the transportation requirements, the light, lean, lethal AEF depends upon rapid and reliable resupply capability, and many modes of transportation are called upon to move all the assets required to sustain an operation. In addition to Air Force aircraft, the Air Force contracted commercial airlift and land transportation, and used sealift. The transportation system was complex and involved coordination among services and among coalition partners.

Part of the transportation system involves distributing goods and assets within the theater. The theater distribution system (TDS) has two components: one that moves initial deployment and sustainment items to where they are needed, from the FSLs to the FOLs, storing many of them in or near the AOR. The second component, a tactical distribution system, provides the onward movement of resources received from CONUS and moves reparable parts to and from FSLs.

The intratheater distribution system appeared to be better organized in OIF than in OEF. Standard air routes were established before combat operations began and adequate airlift was allocated to the AOR for meeting airlift requirements.

TDS was established early and, on the surface, appeared to function well. However, the theater movements system was not always well coordinated with the strategic movements system. Illustrations of gaps between the two systems are the cargo that built up at transshipment points and the problems identifying priorities among services. There continued to be problems in establishing in-transit visibility when shipments moved from one system to the other—from the strategic system to TDS.

Differences in systems for paying for the transportation/shipments also caused problems in the theater distribution system. Strategic airlift draws from an industrial fund on a "pay-as-used" basis, whereas TDS air shipments are free to the shippers. Commercial trucks contracted for TDS use must be paid for on an as-used basis. Moreover, the difference in the pricing of services can cause air assets to be misallocated. For example, since air shipment is "free" (paid out of contingency funds), some cargo that may be better moved by surface transportation (truck) may be delivered by air.[10] This problem arose after major combat operations were over; however, it reflects a systemic problem with the theater distribution system.

Implications (see pp. 89–90)

If the Air Force is responsible for TDS, as it was during JTF NA and OEF, or even if it just provides input to another service that controls TDS, as it did during OIF, the Air Force needs to provide education and training to handle the TDS responsibilities. Creation of a logistics readiness officer can help fulfill this critical need. However, a specific education and training plan for theater distribution needs to be developed.

The transportation system used during any operation will be complex, multimodal, and involve numerous customers (for example, Army, coalition, and Air Force). Theater distribution is more than just the onward movement of spare parts using airlift. The system also includes a network to link forward support locations and CONUS support locations to forward operating locations. MAJCOM components need to work with U.S. Transportation Command (USTRANSCOM) to develop integrated plans to transition peacetime operations smoothly into wartime operations.

---

[10] Telephone interview by Dr. Robert Tripp of Maj Gen Robert Elder, Central Command, Deputy CFACC, August 20, 2003.

An expeditionary Air Force cannot allow critical resources to sit backlogged at FSLs and transshipment points.

Options in having a single party develop an end-to-end military system instead of a strategic movements system and a TDS need to be explored.[11] The distinction between a strategic movements system and a tactical movements system is blurred. For instance, is a system that connects CIRFs or supply FSLs located in one AOR to FOLs in another AOR (as happened with CIRF shipments and other supplies in both OEF and OIF) a strategic system or a tactical system? If it is tactical, which combatant commander should set up the inter-AOR system, the supporting commander or the supported commander? Perhaps the separation of the TDS and the strategic movements system has outlived its usefulness, given the global war on terrorism and the global positioning of combat support resources to meet commitments across a wide variety of scenarios. A review and reconciliation of pricing issues associated with differing shipping modes and continuing efforts to improve in-transit visibility are also needed.

Exploitation of Technology

Findings (see pp. 91–97)

The communications system in place during OIF was much better than the system in place during OEF. The creation of a UTC for communications Air Force Engineering and Technical Service (AFETS) personnel and an Engineering and Technical Service (ETS) program office aided in the deconfliction of taskings and enabled the rapid deployment of taskings to meet changing mission needs.

The theaterwide communication plan that was developed included redundant circuits to most locations, a communication bandwidth increase of nearly 600 percent, and an increase in satellite

---

[11] The Secretary of Defense named USTRANSCOM the Department of Defense Distribution Process Owner on September 16, 2003. USTRANSCOM is responsible for synchronizing global and theater distribution processes.

communications terminals of over 550 percent. The additional bandwidth enabled intelligence data feeds from Global Hawk and Predator to CONUS. The extensive use of precision-guided munitions improved the Air Force's ability to hit suspected targets; improvements in targeting and positioning systems enabled such munitions to be used in any weather.

Not all technology was updated, however. Air Force bare base fuels assets use outdated technology—a combination of pumps, filters, and valves that are not interchangeable and do not employ readily available commercial automation—and require extensive parts and personnel to set them up and operate them.

Implications (see pp. 97–98)

Communications support requirements are no longer limited to just basic bare base systems of local area networks (LANs) and telephone lines. Communications personnel are expected to understand systems and programs. Education and training on operating and maintaining command and control systems need to be developed for communications personnel.

Technological advances in communications and munitions have changed combat support requirements. With personnel in CONUS controlling the flight of unmanned aerial vehicles in the AOR, fewer communications and analysis personnel are required to be deployed forward during an operation. Deploying fewer personnel forward could change the functions of the COMAFFOR Forward and Rear.

The past two operations, OEF and OIF, used precision-guided munitions more often than did JTF NA. Often, fewer smart bombs than dumb bombs are required to achieve a target. Using fewer munitions means a smaller deployment footprint, both in terms of the bombs themselves and in the associated support equipment and personnel.

Fuels is one area in which technology has not been exploited. Each service has different equipment, different training, and different reporting. Better configuration control and interoperability in maintaining bare base fuels assets could reduce both the logistics and personnel footprints. Reducing the number of personnel and amount of equipment taken forward to support the warfighter results in a corresponding reduction in the number of services, security forces, etc.

Resourcing to Meet Contingency, Rotational, and MRC Requirements

Findings (see pp. 99–105)

Analysis of resource usages in the past three operations relative to defense guidance and wartime planning factors indicates that the usage factors associated with supporting permanent rotational commitments and unanticipated contingency operations are different from those used to make programming decisions to obtain resources. An implicit assumption in programming for combat support resources is that the resources necessary to meet major regional conflict (MRC) engagements will cover those needed to support permanent rotations or other contingency operations. We show that these assumptions are not correct and that key combat support resources are stretched thin. The current combat support system and programmed resource base has difficulty simultaneously supporting small-scale contingencies and current rotational deployment requirements. Current usage patterns consume war reserves that may have been planned for use in MRCs. All three past operations posed demands different from those assumed in wartime planning factors. In some cases, the operations have placed as much, or greater, stress on combat support resources as an MRC.

During OIF, some of the shortfalls were alleviated when additional combat support resources were obtained. Additional contract dollars were applied to critical shortages.

The AEF model used to allocate combat support resources is dented, if not severely broken. Before OIF even began, shortages in combat support assets, particularly in high-demand, low-density areas, such as force protection, civil engineering, combat communications, and fuels, stressed the AEF construct, resulting in the Air Force's borrowing against future AEFs during OIF. The current AEF scheduling rules, which allow personnel to be eligible for deployment for only 90 days in a 15-month cycle, were violated (Barthold, 2002, p. 19). Then, in December 2002, the AEF rotation was frozen altogether. Personnel in high-demand/low-density career fields were to remain in the AOR indefinitely.

The management of the expeditionary combat support portion of the deployment was given to the AEF Center. During OIF, the AEF Center found itself sourcing units for deployment. While this responsibility may not have part of the original design, the AEF Center handled it smoothly and was well suited for the job. However, the AEF Center did not source the equipment, the actual aircraft, or the associated flying squadrons.

Implications (see pp. 106–108)

Our findings in three recent operations indicate that the current resource-planning factors and methods are not aligned with current resource-consumption factors. Combat support resources are stretched thin in meeting current rotational, peacekeeping, and training requirements and may leave little capability for meeting future small-scale contingencies, much less potential MRCs. We show that small-scale contingencies such as JTF NA, OEF, and OIF may not necessarily require fewer support resources than an MRC. Actual resource-usage patterns differ from those used in MRC planning computations and, in some cases, small-scale contingencies may require as many resources as MRCs, or even more.

One possible solution would be to change planning factors, increasing the inventory levels of materiel and adding personnel. Computations could be made to determine requirements as a function of the current combat support posture and policies. But with many competing needs, the Air Force may not be able to afford that approach. Several options and trade-offs are available among alternative requirements, alternative combat support distribution options, and other support policies; they may be able to satisfy operational requirements more effectively than just increasing the size of existing pipelines, assuming the current way of providing combat support is the best way.

One such option would be for the Air Force to make investments to decrease delivery time by positioning items closer to the point of need so that they would be distributed to more FSLs in various AORs. Another way to decrease delivery time would be to improve throughput capability of existing FSLs and associated distribution capability. Distribution improvements could be made by increasing working maximum on ground (MOG) at FSL sites or nearby airports, or by improving rail- or sea-handling capabilities. Additional ships to store and move WRM might improve delivery times to FOLs. Smaller, faster ships carrying high-demand assets may help to alleviate some initial airlift concerns. An integrated analysis of options is needed.

WRM requirements and distribution need to be considered jointly. Alternatives to stockpiling munitions and other WRM assets need to be considered in today's uncertain world. One approach might include flexible munitions production with surge capabilities, beyond stocks needed to support the initial phases of likely contingencies.

To evaluate combat support options today requires a capabilities-based assessment of support required in a wide variety of scenarios. A capabilities view of resources is a more appropriate way than a scenarios-based view to consider resource investments in today's world. Using this view, various investments would be stated in terms of what they could support—for example, the ability to support X permanent rotations, a small-scale contingency of Y size (defined by beddown sites), and an MRC of Z size (defined by beddown sites). The Air Force cannot know what scenarios it may be expected to support in the future, but it should be able to state what capabilities it can support.

Systems and organizations need to be developed or refined to enhance expeditionary operations. The AEF Center could have the sole responsibility for nominating all AEF forces for deployment to include aviation UTCs. While this responsibility may not have been part of the AEF Center's original design, the Center handled it smoothly during OIF and is well suited for the job. It did not, however, source much of the equipment, the actual aircraft, or the associated flying squadrons. If these functions were moved from Air Combat Command, Director of Air and Space Operations (ACC/DO) to the AEF Center, and if it were given tools and personnel to manage the equipment issues, it could manage all aspects of the deployment nomination process.

Conclusions (see pp. 109–114)

Combat support execution planning and control processes and command and control organizational alignments have improved since JTF NA and OEF. The implementation of the TO-BE operational architecture has aided in this improvement. Integrating deliberate planning processes and crisis-planning activities requires more work. Although deliberate planning is time-consuming, the process fosters an understanding of the area of responsibility and helps to identify shortfalls. During crisis action planning, there is no time to do the detailed analysis and coordination required during the deliberate-planning stage. Planners should receive training in deliberate planning so that they are prepared for deliberate and crisis action planning.

Austere forward operating locations and an immature theater infrastructure make early planning, knowledge of the theater, and FOL preparation more important. The Air Force recognizes the need to develop these processes and has taken steps to improve them. Survey information to develop FOLs was more readily available during OIF than during the other two operations because of other ongoing operations in the region. Host-nation support was difficult to negotiate, and resultant deployment timelines varied widely throughout the theater.

The current AEF force structure of light, lean, and lethal response forces is highly dependent upon the capacities of forward support locations and throughput. Austere FOLs and immature theater infrastructure illustrate the importance of using FSLs efficiently. Improvements have been made in linking forward support locations and CONUS support locations to dynamic warfighter needs. Much more can be done in this area.

AEF operational goals are dependent upon assured and reliable end-to-end deployment and distribution capabilities that can be configured quickly to connect the selected sets of FOLs, FSLs, and CSLs in contingency operations. By using centralized intermediate repair facilities, the Air Force has traded early strategic lift requirements, used to stock parts at forward operating locations, for a continuous sustainment requirement, to supply the CIRFs. Centralized intermediate repair facilities and other forward support have enabled the combatant commander to deploy more warfighting forces instead of combat support capabilities. However, the continued success of CIRFs relies on dependable resupply, which involves the theater distribution system.

The Air Force may be the predominant user of the theater distribution system in early phases of future campaigns; therefore, the Air Force may be delegated the TDS responsibility. Even if another military service is delegated TDS responsibilities, the Air Force should play an active role in determining TDS capacities and capabilities. The Air Force has made advances in the use of centralized maintenance, expanding its dependence on support from forward support locations. Yet, it finds itself poorly prepared to estimate lift requirements.

Current doctrine splits the responsibility for developing the end-to-end deployment and resupply system among multiple parties, placing the responsibility for developing the strategic movements system on USTRANSCOM and that for intratheater lift on the combatant commander for the AOR. Having one AOR's combat support facilities supporting another AOR's combatant commander—for instance, moving WRM or repaired spares from the European Command AOR to the Central Command AOR—confuses TDS and strategic movements. Where these two systems came together at transshipment points, significant backlogs and system disconnects occurred. This joint doctrine may be inappropriate for expeditionary forces that rely on fast deployment, immediate employment, and responsive resupply of lean, forward-deployed forces. The Air Force's reliance on lean deployments and responsive resupply of deployed units places great importance on the rapid development of contingency end-to-end deployment and distribution capabilities.

During OIF, significant improvements in communications were achieved. Near-real-time raw intelligence data were received in CONUS, then the data were exploited and redistributed to numerous agencies. At the same time, personnel were identifying emerging targets and coordinating attacks—all from inside CONUS. These communications advances reduced the number of expeditionary combat support personnel and equipment deployed; more resources were kept in the rear. There may be other opportunities to exploit technology.

Finally, the planning factors and assumptions that are used to determine resource requirements differ significantly from those that are encountered in current rotational and contingency operations. In many cases, the current resource employment factors are more demanding than the assumptions used to fund resources. This imbalance creates resource shortages that appear in contingency operations. Shortages in combat support assets, particularly in high-demand/low-density areas, such as combat communications, civil engineers (CE), and force protection, stressed the AEF construct.

In addition, the current AEF employment practices differ significantly from planning factors used in the Program Objective Memorandum process to provide for combat support resources. The current AEF scheduling rules are routinely violated in stressed combat support areas. Current AEF scheduling rules may be an effective and efficient means of scheduling and deploying aircraft and aircraft support units; however, the current rules may not be the best for scheduling combat support. Specifically, balances should be maintained between home-station support disruption and deployment commitments.

Below is a list of the recommendations derived from the work on this study. These recommendations are suggested methods to improve agile combat support for the AEF.

Combat Support Execution Planning and Control

- Integrate deliberate planning and crisis planning activities.
- Consider the requirements of both joint and unified commands and identify how to meet those requirements while remaining responsive and adaptive.

Forward Operating Locations and Site Preparation

- Focus attention on political agreements and engagement policies.
- Standardize site-survey procedures and processes within the Air Force, with U.S. allies, and with other services.

Forward Support Location/CONUS Support Location Preparation for Meeting Uncertain FOL Requirements

- Further develop the existing global network of FSLs and CSLs.
- Continue improvements in linking FSLs and CSLs to dynamic warfighter needs.

Reliable Transportation to Meet FOL Needs

- Ensure dependable resupply to CIRFs.
- Identify lift requirements, including airlift, sealift, and movement by land, for theater distribution system.
- Review joint doctrine on the transportation system.
  - —Consider having USTRANSCOM develop end-to-end distribution channel capabilities.
  - —Consider ways to improve TDS performance, including examining pricing mechanisms, and instituting better in-transit visibility and demand-forecasting mechanisms.

- Provide additional training and enhance personnel development policies for the Air Force to meet future theater distribution responsibilities, such as in the exercise EAGLE FLAG.

Exploitation of Technology

- Review contingency combat support functions that could be done in the rear (CONUS)—for example, sustainment planning and execution—because of advances in communications technology that offer the possibility of reducing the forward-deployed footprint.

Resourcing to Meet Contingency, Rotational, and MRC Requirements

- Reevaluate current processes and policies for AEF assignments and the current Program Objective Memorandum assumptions with respect to combat support resources.
- Evaluate existing scheduling rules for combat support with respect to impacts on home-station and deployed combat support performance.

Acknowledgments

Many individuals inside and outside the Air Force provided valuable assistance and support to our work. We thank General Hal Hornburg, Commander, Air Combat Command (ACC/CC), during Operation Iraqi Freedom (OIF), for suggesting that this analysis be conducted. We also thank Lieutenant General Michael Zettler, Deputy Chief of Staff, Installations and Logistics (AF/IL), during OIF for his continuing support of this effort and all our combat support–related research.

We are especially grateful for the assistance of Major General Robert Elder and Colonel Duane Jones. General Elder was the Vice Commander for U.S. Air Forces, Central Command (CENTAF). Colonel Jones is the Director of Logistics for CENTAF, Shaw AFB, South Carolina. Colonel Jones was responsible for planning and executing all combat support operations for U.S. aerospace forces in the theater. Both General Elder and Colonel Jones provided free and open access to everyone under their commands during our search for lessons that might affect the AEF combat support system of the future.

This research would not have been possible without the help and support of the ACC staff. In particular, we thank Brigadier General Patrick Burns, ACC/CE; Brigadier General Michael Collings, ACC/LG; Brigadier General William Lord, ACC/SC; and their staffs. At the AEF Center, we thank Brigadier General Anthony Przybyslawski and his staff. ACC and the AEF Center provided many of the data used in this report.

At the Air Staff, we thank Ms. Susan O'Neal, AF/IL; Mr. Michael Aimone, AF/ILG; Colonel Connie Morrow, AF/ILGX; and their staffs for their support and critique of this work. Each of these people played a key role in the Air Staff Combat Support Center during OIF, and their insights have been instrumental in shaping this research. We also thank Colonel Frank Gorman, Lieutenant Colonel Clifford Fisher, and Lieutenant Colonel William McKinley, HQ USAF/IL Agile Combat Expeditionary Support (ACES), for all their help gathering data and answering questions.

We have enjoyed support for our research from the Air Force's major commands (MAJCOMs) that were involved in OIF. In addition to those mentioned above, we thank Major General Terry Gabreski, AFMC/LG, and Colonel Andrew Busch AFMC/LG/CD, and their staffs. We are especially grateful to Christopher Arzberger, AFMC/LSO, for providing detailed data on spares and transportation.

Finally, at RAND, we thank: Edward Chan, Marc Robbins, Mahyar Amouzegar, and MSgt Les Dishman for their contributions and critiques of our work. We also thank RAND colleagues John Bondanella and Bruce Pirnie for providing thoughtful reviews and critiques of our work. We would also like to thank Mechelle Wilkins, Leslie Thornton, and Dahlia Lichter for their assistance in the many revisions of this report.

Abbreviations and Acronyms

A-4	Logistics Directorate of the Air Component Staff (A-Staff)
AB	Air Base
ACC	Air Combat Command
ACC/CC	Commander, Air Combat Command
ACC/DO	Air Combat Command, Director of Air and Space Operations
ACC/SC	Air Combat Command, Director of Communications and Information
ACES	Agile Combat Expeditionary Support
ACP	Ammunition Control Point
ACS	agile combat support
AEF	air and space expeditionary force
AEFC	Aerospace Expeditionary Force Center
AETF	Air and Space Expeditionary Task Force
AF/IL	Air Force Deputy Chief of Staff for Installations and Logistics
AF/XOX	Deputy Chief of Staff for Air and Space Operations, Director of Operational Plans
AFB	Air Force Base
AFCAP	Air Force Contract Augmentation Program
AFETS	Air Force Engineering and Technical Service

AFFOR	Air Force Forces
AFMC	Air Force Materiel Command
AFSC	Air Force Specialty Code
AFSOC	Air Force Special Operations Command
AGE	aerospace ground equipment
AIS	avionics intermediate-maintenance shop
ALOC	Air Logistics Operations Center
AMC	Air Mobility Command
AMD	Air Mobility Division
ANG	Air National Guard
AOC	Air Operations Center
AOG	Air Operations Group
AOR	area of responsibility
APF	Afloat Prepositioning Fleet
ASETF–NA/CC	Commander, Air and Space Expeditionary Task Force–Noble Anvil
ATO	Air Tasking Order
AWACS	Airborne Warning and Control System
BCAT	Beddown Capability Assessment Tool
C2	command and control
C2ISR	Command and Control Intelligence, Surveillance, and Reconnaissance
CAOC	Combined Air and Space Operations Center
CAT	Contingency Action Team
CC	commander
CCP	Commodity Control Point
CE	Civil Engineer
CENTAF	U.S. Air Forces, Central Command
CENTCOM	Central Command

CFACC	Combined Forces Air Component Command
CHPMSK	Contingency High-Priority Mission Support Kits
CIRF	centralized intermediate repair facility
COMACC	Commander, Air Combat Command
COMAFFOR	Commander, Air Force Forces
COMJTF-NA	Commander, Joint Task Force Noble Anvil
COMJTF-OEF	Commander, Joint Task Force Operation Enduring Freedom
COMUSAFE	Commander, U.S. Air Forces, Europe
CONOPS	Concept of Operations
CONUS	continental United States
CRAF	Civil Reserve Air Fleet
CSAF	Chief of Staff of the Air Force
CSC	Contingency Support Center
CSC2	combat support execution planning and control
CSL	CONUS support location
CWT	Customer Wait Time
DATCALS	Deployable Air Traffic Control and Landing Systems
DCC	Deployment Control Center
DDOC	Deployment/Distribution Operations Center
DEPORD	Deployment Order
DGATES	Deployable Global Air Transportation Execution System
DIRMOBFOR	Director of Mobility Forces
DISA	Defense Information Systems Agency
DLA	Defense Logistics Agency

DoD	Department of Defense
ECM	electronic countermeasure
ECS	expandable common-use shelter; Expeditionary Combat Support
EETL	Estimated Extended Tour Length
ETS	Engineering and Technical Service
EUCOM	European Command
EW	Electronic Warfare
FAM	functional area manager
FAO	Foreign Area Officer
FMSE	fuels mobility support equipment
FOC	fully operationally capable
FOL	forward operating location
FSL	forward support location
GAT	Global Assessment Team (AMC)
GATES	Global Air Transportation Execution System
GCS	Ground Control Station
GPS	Global Positioning System
HIT	High Impact Target
HQ	Headquarters
HUMRO	humanitarian relief operation
ICP	Inventory Control Point
IOC	initially operationally capable
ISR	intelligence, surveillance, and reconnaissance
ITV	in-transit visibility
JAOC	Joint Air and Space Operations Center
JAOP	Joint Air Operations Plan
JCS	Joint Chief of Staff
JDAM	Joint Direct Attack Munition

JEIM	Jet Engine Intermediate Maintenance
JFACC	Joint Forces Air Component Commander
JMC	Joint Movement Center
JOPES	Joint Operations Planning and Execution System
JPO	Joint Petroleum Office
JTF NA	Joint Task Force Noble Anvil
K2	Karshi Khanabad
KU	Kuwait; a bandwidth
LAN	local area network
LANTIRN	Low Altitude Navigation and Targeting Infrared for Night
LG	Logistics
LRU	line replaceable unit
LSA	Logistics Sustainability Analysis
LSO	Logistics Support Office
MAAP	Master Air Attack Plan
MAJCOM	major command
MHE	material handling equipment
MILAIR	military airlift
MOE	Measure of Effectiveness
MOG	maximum on ground
MRC	major regional conflict; major regional contingency
MRE	meals, ready-to-eat
MX	Maintenance
NAF	numbered Air Force
OAF	Operation Allied Force
OEF	Operation Enduring Freedom
OIF	Operation Iraqi Freedom

ONE	Operation Noble Eagle
OSC	Operations Support Center
PACAF	Pacific Air Forces
PDC	Power Distribution Center
PGM	precision-guided munition
POD	port of debarkation
POE	port of embarkation
POL	petroleum, oil, and lubricants
POM	Program Objective Memorandum
Prime BEEF	Priority Improved Management Effort–Base Engineer Emergency Force
PSAB	Prince Sultan Air Base
RAF	Royal Air Force
RED HORSE	rapid engineer deployable heavy operational repair squadron engineer
REPOL	report on petroleum, oil, and lubricants
RFF	Request for Forces
ROE	rules of engagement
RS	receive suite
RSS	Regional Supply Squadron
SATCOM	satellite communications
SDMI	Strategic Distribution Management Initiative
SITREPS	situation reports
SOF	Special Operations Forces
STAMP	Standard Air Munitions Package
STAR	standard air route
STEP	Survey Tool for Employment Planning
SWA	Southwest Asia
TALCE	Tactical Airlift Control Element

TAV	Total Asset Visibility
TBMCS	Theater Battle Management Core System
TDS	theater distribution system
TFEL	Task Force Enduring Look
TPFDD	Time Phased Force and Deployment Data
USAFE	U.S. Air Forces, Europe
USCOMEUCOM	Commander, European Command
USTRANSCOM	U.S. Transportation Command
UTASC	USAFE Theater Aerospace Support Center
UTC	unit type code
WCMD	Wind Corrected Munitions Dispenser
WOC	Wing Operations Center
WRM	war reserve materiel
WSD	Warfighter Sustainment Division
WWX	World Wide Express

CHAPTER ONE

Introduction

The Air and Space Expeditionary Force (AEF) concept—substituting speed of deployment and employment for presence—was developed to allow the Air Force to respond quickly, with a tailored, sustainable force, to any national security issue. Since 1997, RAND Project AIR FORCE and the Air Force Logistics Management Agency have studied options for configuring a future Agile Combat Support (ACS) system that would enable AEF goals to be achieved (Galway et al., 2000; Tripp et al., 1999).

Background of the Agile Combat Support System

As defined in Tripp et al. (2000), AEF operational goals are to

- foster an expeditionary—forward-thinking—mind-set
- rapidly configure support needed for the forces selected to achieve the desired operational effects
- quickly deploy both large and small tailored force packages with the capability to deliver substantial firepower anywhere in the world
- immediately employ these forces upon arrival
- smoothly shift from deployment to operational sustainment
- meet the demands of small-scale contingencies and peacekeeping commitments while maintaining readiness for potential contingencies outlined in defense guidance.

Two earlier RAND studies (Galway et al., 2000; Tripp et al., 1999) present the framework for an ACS system to support the AEF concept:

- An expeditionary mind-set that has been instilled in combat support personnel
- A combat support execution planning and control (CSC2) system to assess, organize, and direct combat support[1] activities, meet operational requirements, and be responsive to rapidly changing circumstances. The CSC2 capability would help combat support personnel
  - —estimate combat support resource requirements and process performances needed to achieve the desired operational effects for the specific scenario.
  - —configure supply chains for deployment and sustainment, including the military and commercial transportation needed to meet deployment and sustainment needs.
  - —establish control parameters (for example, goals, maximum or minimum requirements) for the performance of various combat support processes required to meet specific operational needs.
  - —track actual combat support performance against control parameters.
  - —signal when a process is outside accepted control parameters so that plans can be developed to get the process back within control limits.
- A quickly configured and responsive distribution network to connect forward operating locations (FOLs), forward support locations (FSLs), and continental United States (CONUS) support locations (CSLs)

---

[1] In this report, the term *combat support* is defined as anything other than the actual flying operation. Combat support consists of civil engineering, communications, security forces, maintenance, service, munitions, etc. Not all aspects of combat support are addressed in this report because the scope was too broad.

- A network of FOLs resourced to support varying deployment/employment timelines
- A network of FSLs configured outside CONUS to provide storage capabilities for heavy war reserve materiel (WRM), such as munitions and tents, and selected maintenance capabilities, such as centralized intermediate repair facilities (CIRFs) that service jet engines of units deployed to FOLs. FSLs could be collocated with transportation hubs.
- A network of CSLs, including Air Force depots, CIRFs, and contractor support facilities. As with FSLs, a variety of different activities may be set up at major Air Force bases, convenient civilian transportation hubs, or Air Force or other defense repair depots (Tripp et al., 1999).

Figure 1.1 is a notional representation of how the basic ACS concepts are being integrated to form a global ACS network that can enable AEF operational goals across a wide variety of scenarios, in areas where an operation might likely occur.

The ACS Network in Joint Task Force Noble Anvil, Operation Enduring Freedom, and Operation Iraqi Freedom

The ACS framework was designed to support a wide variety of operational scenarios, from small-scale contingencies to major regional conflicts. In this report, we evaluate three recent U.S. military operations: JTF NA, OEF, and OIF. These operations differed significantly; therefore, experiences with the ACS network concepts in these different contingencies should be of substantial interest to ACS concept developers. The ACS network is evolving continuously. Several elements of the ACS framework were implemented before Operation Iraqi Freedom (OIF), in support of Operation Enduring Freedom

Figure 1.1
Conceptual Global ACS Network

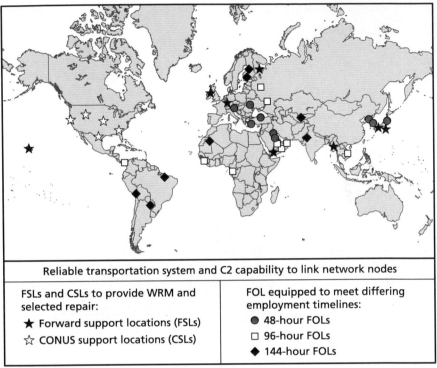

Reliable transportation system and C2 capability to link network nodes	
FSLs and CSLs to provide WRM and selected repair: ★ Forward support locations (FSLs) ☆ CONUS support locations (CSLs)	FOL equipped to meet differing employment timelines: ● 48-hour FOLs □ 96-hour FOLs ◆ 144-hour FOLs

RAND *MG193-1.1*

(OEF) in Afghanistan and Joint Task Force Noble Anvil (JTF NA)[2] in Serbia. Other elements were implemented during OIF. JTF NA, OEF, and OIF provide three important opportunities to study how AEF ACS concepts were implemented during contingency operations and how they have been refined with each contingency to better support AEF goals.[3]

Differing degrees of the ACS network were implemented to enable operations in all three contingencies (see Figure 1.2). Some of

---

[2] The U.S. portion of Operation Allied Force was code-named Joint Task Force Noble Anvil. This report concentrates on Air Force operations conducted by JTF NA.

[3] This report does not address medical support issues.

the ACS concepts were either not fully developed or lacked understanding by the combat support and operational community before
JTF NA and OEF began. For instance, both JTF NA and OEF contingency operations provided the opportunity to study how CSC2
concepts were implemented in an ad hoc manner, without following
specific doctrine. OIF allowed analysis of how the future, or "TO-
BE," CSC2 operational architecture affected the CSC2 development.

In all three contingencies, the Air Force played a major role in
the development of the theater distribution system (TDS). Both
CSC2 and TDS development were problems in JTF NA, and they
continued to be problems in OEF. However, much was learned from

Figure 1.2
ACS Network as Implemented During JTF NA, OEF, and OIF

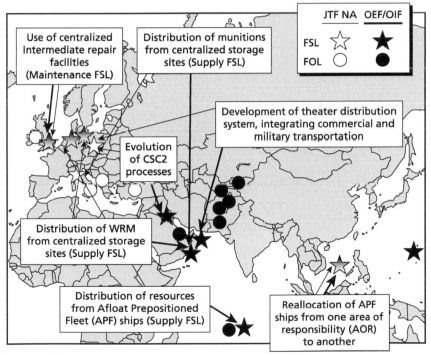

these experiences, allowing a systematic and effective development of CSC2 and improved TDS during OIF.

In all three operations, forward operating locations were opened that had not been used in previous Air Force exercises or operations. Forward support locations were used extensively in all three operations. In the case of maintenance FSLs, much was learned during JTF NA. That knowledge was successfully transferred to personnel engaged in OEF and OIF. Maintenance FSLs were used to repair fighter jet engines, Low Altitude Navigation and Targeting Infrared for Night (LANTIRN), electronic warfare pods, and F-15 avionics components. In OEF and OIF, U.S. Air Forces, Europe (USAFE) maintenance FSLs supported operations in the U.S. Air Forces, Central Command (CENTAF) AOR. WRM or supply FSLs were extensively used during all three operations, including the use of the Afloat Prepositioning Fleet (APF; FSLs afloat). Experiences with these concepts can help improve their implementation. Concentrating only on Air Force operations, this report provides an analysis of ACS activities during OIF and compares those activities with activities during JTF NA and OEF. All three operations are compared with the concepts designed to support the AEF. Specifically, this report analyzes the implementation of ACS concepts during these contingencies to determine (1) whether these concepts should be modified and (2) whether these three recent experiences presented additional ACS areas that need to be addressed.

Analytic Approach

Evaluating the performance of combat support in any operation raises the question: What measures should be used to judge how well the operation was supported? Evaluation could be based on whether constraints on combat support inhibited operations during a contingency operation. For example, were sorties not flown because combat support was lacking or did units not meet operational missions because they did not have the needed support? In all three operations, JTF NA, OEF, and OIF, combat support did not inhibit operational mission performance. Thus, on the surface, it may appear that there may be little to learn from these operations.

In this analysis, as in the JTF NA and OEF studies, the focus is on system performance relative to the ACS concepts that were developed to enable AEF goals. Official accounts of JTF NA indicate that "the logistics system . . . [and] logistics support [were considered a] success story during the air war" (Headquarters, USAF, 2000, pp. 45–46). Yet, bottlenecks and analysis of scarce resources indicate several areas in which improvement may be necessary to better attain AEF goals.

Statements from both OEF operational and combat support leaders indicate that significant combat support issues were associated with OEF that raise serious concerns about supporting future contingencies. In the April 2002 issue of *Armed Forces Journal International,* Lt Gen Michael E. Zettler, Air Force Deputy Chief of Staff for Installations and Logistics (AF/IL) during OEF, said "we made it up for Afghanistan as we went along . . . every one of those [missions] was an opportunity for failure . . . everything is needed" in that region of the world. Maj Gen Jeffrey B. Kohler, Director of Operational Plans for the Deputy Chief of Staff for Air and Space Operations (AF/XOX) during OEF, in a discussion in December 2001, expressed concern that in an operation in which comparatively little "iron" was pushed forward, combat support resources were surprisingly stressed.

Statements from both operational and combat support leaders indicated that there were great improvements during OIF. In April 2003, Gen Hal Hornburg, Commander, Air Combat Command (ACC/CC), said "we need to capture the magnificent success we had deploying . . . I suspect we have data (and memories) of past experiences which did not go nearly as well." Brigadier General Patrick Burns, Commander, Air Force Forces (COMAFFOR)/A-7 during OIF, praised the ability of combat support personnel to understand and express requirements in terms of their operational effects. For example, "The new ramp will give CENTAF parking for 4 more B-1s or 152 more JDAMs [Joint Direct Attack Munitions] per day."

In this report, we use both empirical data and interview data from numerous sources to analyze the following ACS areas: the development of combat support execution planning and control, the preparation and development of forward operating locations, the preparation of forward support locations/CONUS support locations, reliable transportation to meet the needs of forward operating locations, exploitation of current technology, and resourcing to meet the requirements of contingency, rotational, and major regional conflict (MRC) operations.

To gather information on the implementation of CSC2 concepts, we discussed the combat support chain of command with key participants, including the COMAFFOR A-4, the COMAFFOR A-4 Rear, CENTAF/CV, Commander, USAFE, Theater Aerospace Support Center (UTASC/CC), USAFE/LG, AF/IL, and their staffs. We gathered JTF NA data during extensive field interviews,[4] from Air Force publications, and from Joint doctrine. We collected OEF and OIF data from situation reports (SITREPS), Air Force Combat Support Center daily briefings, CENTAF computer sites, and ACC Contingency Action Team (CAT) and Air Force Forces (AFFOR) A-4 Rear computer sites.

For JTF NA, USAFE/LG, numerous CONUS-based Air Force organizations, and the Operation Allied Force (OAF) Time Phased Force and Deployment Data (TPFDD) provided data on the development of FOLs. For OEF, we gathered information on FOL timelines from Air Force Combat Support Center daily briefings, data from Task Force Enduring Look (TFEL)[5] including SITREPS, and the OEF TPFDD. TFEL also provided data on OEF executive agency responsibility. For OIF, we gathered execution data from the Joint Operations Planning and Execution System (JOPES), Air Force Combat Support Center daily briefings, and field interviews.

---

[4] Interviews were with the following individuals and organizations. In USAFE: LG, LGX, LGM, LGW, LGS, LGT, XP, the Air Mobility Operations Control Center, and the 32nd Air Operations Squadron. In AFMC: ALG and XPS. Other organizations included ACC/XR, PACAF/LG, 3 AF, 16 AF, 7 AF, EUCOM/J-4, PACOM/J-4, USFK/J-4, and personnel at Aviano Air Base (AB), Royal Air Force (RAF) Lakenheath, and Spangdahlem AB.

[5] Formed in October 2001, Task Force Enduring Look is an Air Force–wide data collection, exploitation, documentation, and reporting effort for the air campaign against terrorism.

To gather data on host-nation and contractor support, we conducted numerous field interviews during JTF NA. To obtain data for OEF, we interviewed CENTAF A-4 staff, the Director of Mobility Forces (DIRMOBFOR), and contractors on-site at CENTAF and in the area of responsibility (AOR). For OIF, we gathered information during field interviews with various personnel, including the CENTAF A-4 staff, contractors on-site in the AOR, and staff at the Air Force Combat Support Center.

In analyzing the amount of materiel moved and the method of transportation, we consulted the OAF, OEF, and OIF TPFDDs and/or execution data, as well as data provided by the contractor in the area of responsibility. The Air Force Combat Support Center Fuels personnel provided data on fuels. CENTAF/LGX provided data on munitions and rations. We obtained data on spares from Air Force Materiel Command/Logistics Support Office–Transportation (AFMC/LSO-LOT). To gather information about FSL throughput constraints and TDS, we interviewed the DIRMOBFOR, CENTAF staff, and the Deputy Director of the Joint Movement Center (JMC). Air Mobility Command (AMC) and U.S. Transportation Command (USTRANSCOM) staff were also interviewed.

The Aerospace Expeditionary Force Center (AEF Center) provided information about the current AEF organizational structure for ACS, and CENTAF/LGX provided data on WRM. The AEF Center, the Air Force Combat Support Center, and the Air Staff all provided data about stressed career fields and the effect of resourcing the current AEF construct on those stressed career fields.

In this report, specific experiences from OIF are compared with those experiences documented during JTF NA and OEF. The experiences are compared to determine any similarities or differences. When an experience was different or new, we assessed why it was different. Also, we examined experiences in an attempt to determine whether lessons from JTF NA and OEF had been acted on to improve implementation on ACS concepts during OIF.

Finally, we must point out that our analysis indicates how well the Air Force ACS system performed in OIF, not necessarily how well the system could perform if demands were greater. Understanding experiences from this implementation could be of value for combat support and operations personnel who may be called upon to support future contingency operations.

Organization of This Report

All three operations provide important opportunities to study how ACS concepts were implemented during contingency operations. In Chapter Two, we provide an overview of all three operations, for comparison. We then address six areas: CSC2 structure, in Chapter Three; FOL development, in Chapter Four; the use of FSLs and CSLs, in Chapter Five; the transportation system, in Chapter Six; the use of current technology, in Chapter Seven; and resourcing to meet current operational requirements, in Chapter Eight. Understanding these experiences could be of value for combat support and operational personnel who may be called upon to support future contingency operations. Chapter Nine concludes with a summary of opportunites for improving combat support for the AEF of the future.

CHAPTER TWO

Overview of JTF NA, OEF, and OIF

Every military operation has its own unique characteristics. There-fore, neither the performance of the current support system nor the design of a future combat support system should be judged solely on the basis of any one experience. That said, JTF NA, OEF, and OIF provide important experiences that warrant study. In this chapter, we discuss some of the specific combat support characteristics of JTF NA, OEF, and OIF. We begin with an overview of the size and scope of all three operations. We then discuss the support required to con-duct each of the three operations.

Operations

For comparison, Figure 2.1 presents the size and scope of Joint Task Force Noble Anvil, Operation Enduring Freedom, and Operation Iraqi Freedom. Although considered small-scale, JTF NA, OEF, and OIF are not small in all aspects. The figure provides a quantitative comparison of the approximate number of Air Force sorties flown, amount of munitions expended, number of beddown locations, and number of Air Force personnel and aircraft deployed in recent opera-tions.

Figure 2.1
JTF NA, OEF, and OIF Size and Scope, for Comparison

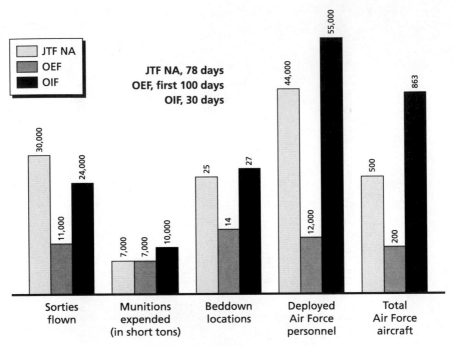

SOURCES: Data for Joint Task Force Noble Anvil were collected from USAFE and CONUS-based Air Force organizations and abstracted from the OAF TPFDD; OEF data are from Maj Robert Rosenthal, "Air Force Operations Group Noble Eagle/Enduring Freedom Operations Update," briefing, January 15, 2002, and CENTAF Munitions Expenditure Rollup, March 15, 2002; data for OIF are from U.S. Air Force, Central Command, *Operation Iraqi Freedom—By the Numbers*, report by CENTAF Assessment and Analysis Division, Prince Sultan Air Base, Saudi Arabia, April 30, 2003.
RAND *MG193-2.1*

Although it would be logical (because of the location in the same AOR) to compare Operation Iraqi Freedom with the last major operation, Operation Enduring Freedom, the size of OIF in sorties flown, 24,000, was closer to that of JTF NA, approximately 30,000 sorties flown (excluding Special Operations Forces [SOF], Army Helo, and coalition sovereignty flights). More munitions were dropped during OIF (10,000 tons) than in either JTF NA or OEF (7,000 tons in each) in a shorter amount of time (30 days) than in

JTF NA (78 days) and OEF (the first 100 days). In Operation Iraqi Freedom, a much higher percentage of the munitions (68 percent of the number, or approximately 75 percent of tonnage) that were dropped were precision-guided, compared with approximately 25 percent of the number in JTF NA and 46 percent of the number in OEF.

The total number of beddown locations used in OIF (27)[1] is approximately the same as those used during JTF NA (25). However, note that only beddown locations within the AOR were counted in these totals. The 27 beddown locations used during OIF were located in 17 different countries. In OEF, the 14 FOLs that were used in the AOR were located in 10 different countries.[2]

During JTF NA, approximately 500 Air Force aircraft and 44,000 personnel were deployed. JTF NA involved a large fighter force consisting mainly of USAFE-based forces with some augmentation from CONUS forces and some bombers. OEF, on the other hand, involved few fighters and a larger number of deployed bombers based out of Diego Garcia, for a total of 200 aircraft and 12,000 personnel. In the buildup for Operation Iraqi Freedom, approximately 900 Air Force aircraft and 55,000 Air Force personnel were deployed to 27 locations (U.S. Air Force, Central Command, 2003). OIF involved a large fighter force from CONUS, combined with units previously deployed to the region in support of Operation Southern Watch and the support of some bombers.

JTF NA required minimal Special Operations Forces, (whereas both OEF and OIF used such forces extensively. Both OEF and OIF involved large naval participation; JFT NA did not. Tankers were used to support operations during JTF NA, and OEF operations relied heavily on tankers for bomber and naval fighter operations, to

---

[1] Beddown locations changed during the course of OIF. The number listed here is an approximation. Operations were still ongoing in Iraq at the time this report was being written. The actual number and location of beddown locations are classified.

[2] In Operation Enduring Freedom, several European Command (EUCOM) bases supported humanitarian airdrop missions, including the associated fighter escorts and tanker requirements.

support strikes within Afghanistan because of the distances from naval- and land-based air strike packages. Similarly to OEF, OIF relied heavily on tankers because of the distances from bases to targets.[3] The operational command and control (C2) of forces during JTF NA was relatively straightforward: All remained within one command, and intelligence, surveillance, and reconnaissance (ISR) and C2 assets were not overly stressed. By contrast, OEF and OIF placed large demands on the Airborne Warning and Control System (AWACS) and other ISR assets, such as U-2s, Predators, and Global Hawks; C2 often crossed major commands (MAJCOMs). JTF NA, as part of Operation Allied Force, was involved with the significant coalition forces participating in Operation Allied Force. Both OEF and OIF had limited coalition participation.

Support Requirements

In terms of the number of aircraft, people, and beddown locations, Operation Iraqi Freedom was larger than JTF NA. Using about the same number of locations (27 as opposed to 25), it had 25 percent more personnel and 80 percent more aircraft. The infrastructure support requirements during OIF were much larger than those for JTF NA: nearly three times the fuels equipment, twice the aerial port operations requirements, and nearly ten times the FOL support requirements (see Figure 2.2), and five times the beddown construction requirements. Relative to the size of the operation, OIF requirements were more comparable to OEF requirements. Fuels and aerial ports requirements in OIF were somewhere between two and three times the corresponding requirements of OEF, for an operation that had approximately four times the aircraft and personnel.

---

[3] Although this report does not address other operations, it is important to note that the large number of tankers deployed in support of OIF could have had a potentially detrimental effect on homeland security: An increased alert level in CONUS would have significantly stressed the remaining tanker force.

Figure 2.2
Combat Support Requirements

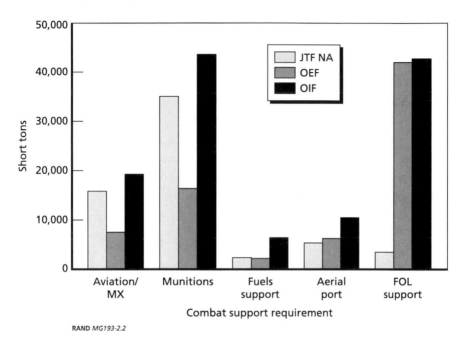

RAND MG193-2.2

Interestingly, the total FOL support requirements programmed in the execution plan for OIF was less than for OEF.[4] However, this disparity may be due to OIF's employing many of the same bases as OEF, at which infrastructure development had already occurred. According to CENTAF, of the 27 locations used for Operation Iraqi Freedom, 13 were new, including some airfields seized within Iraqi territory. These new forward operating locations required the use of bare base support assets, including 21 Harvest Falcon housekeeping kits for beddown of personnel (each kit handling approximately 1,100 personnel), 5 Initial Flightline kits and 5 Industrial Operations kits for airfield operations (typically, one of each kit is needed for

---

[4] Not included in the TPFDD were 5 Housekeeping sets, 1 Industrial Operations set, and 1 Initial Flightline set, which were used later in support of OIF.

each new location), and 3 Follow-On Flightline kits for additional aircraft beddown support.

Some other OIF statistics reflecting the size of the combat support effort are shown in Table 2.1. The percentage of growth is a comparison of OIF with Operation Southern Watch/Operation Northern Watch. Most meals and water were initially brought into the theater until contracts could be established to provide these services. The civil engineering improvements and communication upgrades in-place during OIF far exceeded those of any previous operation. The magnitude of the beddown and contingency construction effort included 820,000 square yards of concrete and asphalt airfield pavement laid, 3,200 bare base tents erected, 3.2 million square feet of contingency facilities constructed, 190 miles of expedient water piping installed, and 200 million gallons of fuels storage and distribution built.

Table 2.1
OIF Combat Support Statistics

Vehicles	2,374
Meals served daily	111,000
Water consumed daily (liters)	989,865
Engineering contracts (number)	211
Engineering contracts (U.S. dollars)	$329,000,000
Bandwidth	783 MB (+596%[a])
Commercial satellite communications (SATCOM) terminals	34 (+560%[a])
Average terrestrial bandwidth	10 MB (+444%[a])

[a] Indicated percentage of increase during OIF over Operation Southern Watch/Operation Northern Watch.

Pacific Air Force Flexible Deterrent Option

While forces were being built up in Central Command for Operation Iraqi Freedom, tensions on the Korean peninsula increased. When the Navy aircraft carrier battle group normally stationed in the Pacific Command moved to the Central Command AOR, Air Force fighters and bombers were deployed to fill in the gap and provide an additional deterrent force. Initial planning for this Flexible Deterrent Option began in October 2002, and the Deployment Order was signed in early February 2003. The first movement of personnel and equipment began shortly thereafter. Although planned in advance, the plans for the flexible deterrent option were neither detailed nor accurate.

While waiting for official deployment orders, the Pacific Air Forces (PACAF) attempted to tailor the support packages that were coming to the Pacific Command with the supporting bases. By the time official tasks were received, it was too late to tailor the support; consequently, whole mobility packages were sent to PACAF. For example, Dyess AFB, Texas, and Barksdale AFB, Louisiana, sent two planeloads of unneeded aerospace ground equipment (AGE) to Andersen AB, Guam.[5] ACC was inundated with CENTAF demands and did not work potential PACAF issues.

In early March 2003, 24 bombers were deployed to Guam; 24 additional fighters were deployed to Korea later that month. These deployments required 21 tankers to form the air bridge.[6]

Although, internally in PACAF, the moves went smoothly, they put unanticipated stress on Air Staff and ACC. In addition to the combat aircraft, the following combat support assets had to be moved to Guam as well: Expeditionary Medical Squadron; headquarters staff support for air expeditionary wing; combat communications; aircraft maintenance; intelligence; comptroller; munitions buildup and maintenance; supply augmentation; petroleum, oil, and lubricants (POL)

---

[5] Data are from Pacific Air Force/LG-ALOC (Air Logistics Operations Center), Hickam AFB, Hawaii.

[6] Data are from Pacific Air Force/LG-ALOCX.

augmentation (includes fuel trucks); Office of Special Investigations; vehicle operations and maintenance; weather personnel; security forces; Public Affairs.

The Korean peninsula saw increases in the following combat support assets: material handling equipment operations, contracting, finance, POL augmentation, personnel teams, vehicle operations/maintenance, weather personnel, Office of Special Investigations, services, intelligence, en route support teams, maintenance, satellite communications capability, security forces, civil engineering/ Prime BEEF,[7] and munitions.

OIF in Perspective

OIF provides one glimpse of potential future conflicts. Joint Task Force Noble Anvil planners had several months to plan the operation, and it was a smaller operation conducted from bases with good infrastructure. Operation Enduring Freedom was a small operation, but it was conducted on short notice in an immature theater. Operation Iraqi Freedom was a large operation, sized similarly to a major regional conflict, but it had the benefit of long planning and buildup times. The next conflict could be similar to any one of these recent operations, or it could be an entirely different scenario.

In recent comments by the Chief of Staff of the Air Force (CSAF) Gen John Jumper, "our heightened tempo of operations is likely to continue at its current pace for the foreseeable future." The Air Force should be able to support the deployment of a large number of forces, either all at once in a major conflict or in an accumulation of a number of small-scale contingencies. Furthermore, it should be able to do so on short notice and in austere environments, particularly as the war on terrorists continues around the world.

---

[7] Prime BEEF stands for Priority Improved Management Effort–Base Engineer Emergency Force.

Combat Support Execution Planning and Control

In this chapter, we address some of the key combat support execution planning and control (CSC2) experiences from OIF and compare them to those in JTF NA and OEF. We look at timelines for each operation and the command and control organizational structure employed during each operation. Also presented are CSC2 implications from past experiences for meeting future AEF goals.

CSC2 Nodes and Responsibilities in JTF NA

CSC2 development in JTF NA was accomplished iteratively, over many months. Each new iteration took into consideration a new set of planning factors and operational requirements. Formal planning for the air war over Serbia took place in late 1998. The first round of planning culminated in the creation of Joint Task Force Noble Anvil in January 1999 (see Figure 3.1). Even with many months to plan, CSC2 development was ad hoc and did not follow doctrine.

Doctrine calls for a numbered Air Force (NAF) to transition to the wartime Air Force component role in times of conflict. Doctrine also calls for the augmentation of the NAF for reachback capability. During JTF NA, the Air Force chose to deviate from doctrinal guidelines and separated the AFFOR and Joint Forces Air Component Commander (JFACC) staffs into two separate locations (see

Figure 3.1
JTF NA Operational and CSC2 Timeline

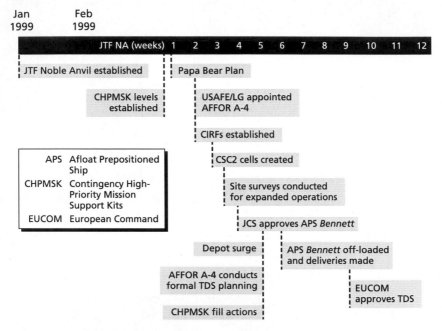

SOURCES: 32nd Air Operations Squadron, Ramstein AB, Germany, transcript of interview, January 31, 2000; Headquarters, USAFE Crisis Action Team, transcript of interview, February 2, 2000.

RAND MG193-3.1

Figure 3.2).[1] Lt Gen Michael Short, 16th Air Force Commander, was selected to be the JFACC. The 16th Air Force A-4[2] was quickly

---

[1] According to Air Force Doctrine Document 2, Organization and Employment of Aerospace Power (U.S. Air Force, 1998), the NAF/CC and staff are delegated the COMAFFOR responsibilities. The NAF commander acts as both the AFFOR, providing all the necessities from tents to food and munitions for Air Force forces, and the JFACC, overseeing the employment of all aerospace forces. Accordingly, the NAF staff is designated as both the AFFOR staff and the JFACC staff, filling the Joint Air and Space Operations Center (JAOC) and, when necessary, manning the JTF staff. Based on their doctrinal responsibilities, the NAF staff is the principal warfighting staff.

Figure 3.2
CSC2 Organizational Structure During JTF NA

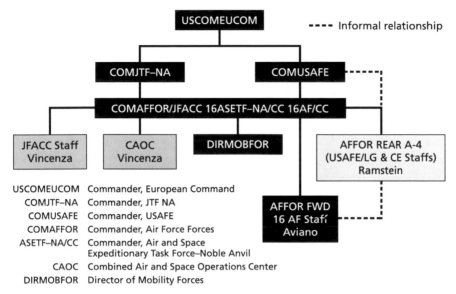

USCOMEUCOM   Commander, European Command
COMJTF–NA   Commander, JTF NA
COMUSAFE   Commander, USAFE
COMAFFOR   Commander, Air Force Forces
ASETF–NA/CC   Commander, Air and Space
                       Expeditionary Task Force–Noble Anvil
CAOC   Combined Air and Space Operations Center
DIRMOBFOR   Director of Mobility Forces

SOURCE: Gen John P. Jumper, "The Limitations of Doctrine," briefing to Air Force
Doctrine Center, Maxwell AFB, Ala., n.d.

RAND *MG193-3.2*

overwhelmed by his responsibilities and looked to the MAJCOM
component, USAFE, to provide support. At the beginning of JTF
NA, USAFE did not have clearly established roles and responsibilities
to execute these contingency responsibilities. The staff faced challenges in organizing to provide this support. They struggled to estimate their support needs and present them to the European Command.[3]

As JTF NA progressed, organizational roles and responsibilities
evolved. JTF NA revealed the need to evaluate systems and processes
to improve CSC2.

---

[2] The term *A-4* refers to the Logistics Directorate of the air component staff, which is responsible for logistics planning and execution for all Air Force activities in the area of responsibility.

[3] Information is based on extensive interviews and data collected from USAFE and
CONUS-based Air Force organizations.

The TO-BE Operational Architecture

As learned in JTF NA, CSC2 processes were not well documented in either Air Force doctrine or joint doctrine. Doctrine outlined general guidelines for command line structure, but it did not clearly specify roles and responsibilities for CSC2 processes. As a result, operational and combat support communities had limited understanding of the CSC2 process. This lack of understanding and an ad hoc organization resulted in problems in combat support command and control in JTF NA.

In response to the CSC2 issues discovered during the air war over Serbia, AF/IL asked RAND Project AIR FORCE to study the current CSC2 operational architecture and develop a future, or "TO-BE," CSC2 operational architecture (Leftwich et al., 2002). The TO-BE architecture would define CSC2 processes and roles and responsibilities associated with those processes. It could be implemented Air Force–wide to standardize command and control for combat support.

In 2000 and 2001, RAND Project AIR FORCE documented the current processes, identified areas in need of change, and developed processes for a well-defined, closed-loop[4] TO-BE CSC2 operational architecture that incorporated the lessons learned during JTF NA.

More specifically, the TO-BE CSC2 operational architecture identifies the future CSC2 functions to include the ability to

- enable the combat support community to quickly estimate combat support requirements for force-package options needed to achieve desired operational effects and assess the feasibility of operational and support plans
- quickly determine beddown capabilities, facilitate rapid TPFDD development, and configure a distribution network to meet employment timelines and resupply needs

---

[4] A *closed-loop process* takes the output and uses it as the input for the next iteration of the process.

- facilitate execution resupply planning and performance monitoring
- determine effects, in-theater as well as globally, of allocating scarce resources to various combatant commanders
- indicate when combat support performance deviates from the desired state and implement replanning and/or get-well planning analysis (Leftwich et al., 2002).

The TO-BE operational architecture was completed in September 2001, just as OEF began. See the Appendix for a list of CSC2 TO-BE nodes and their responsibilities. More details about the CSC2 TO-BE operational architecture can be found in Tripp et al. (2004).

CSC2 Nodes and Responsibilities in OEF

The CSC2 TO-BE architecture was not implemented during OEF because it had not been completely vetted to senior leadership. However, OEF provided an opportunity to review the CSC2 processes and improve the architecture's design.

As in JTF NA, combat support command relationships during OEF did not follow doctrine. Doctrine called for augmenting CENTAF A-4 personnel; elements of the CENTAF A-4 deploying forward, if forward operations were necessary; and reachback A-4 presence at the CENTAF Rear site at Shaw AFB, South Carolina. Instead of augmenting the NAF, the CENTAF A-4 and A-7, with the ACC/LG and ACC/CE, established augmentation arrangements with ACC at Langley AFB, Virginia. Langley supported the A-4 and A-7 reachback responsibilities of the CENTAF A-4/A-7, who went forward to Prince Sultan Air Base in Saudi Arabia to work with the COMAFFOR/JFACC (see Figure 3.3).

At the beginning of OEF, the CENTAF A-6 already had forward and rear elements in place for Operation Southern Watch. These elements were augmented during OEF by activating parts of

Figure 3.3
CSC2 Organizational Structure Implemented During OEF

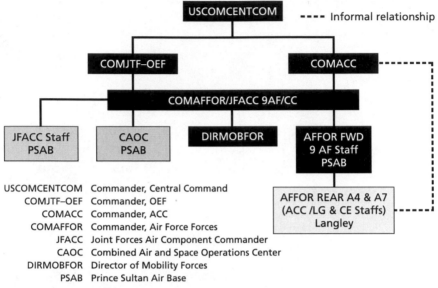

RAND *MG193-3.3*

the Air National Guard associated with CENTAF and with ACC Director of Communications and Information (ACC/SC). The A-6 reachback originally operated from Shaw AFB. A few weeks into the operation, staff relocated to Langley AFB to more closely integrate with the ACC/SC Crisis Action Team Support Cell, but later moved back to Shaw AFB.

Both JTF NA and OEF used differing approaches to the practice of reachback to supporting MAJCOMs. However, in each case, this support was developed at the time combat support was actually being executed. As a result, no functions had been clearly defined or documented for the Rear Staff. Working out the implementation of reachback was left to the particular players who were occupying the various CSC2 positions, without the aid of a playbook. Under such circumstances, effectiveness may reflect the skills and experience of the players more than doctrine and policies.

During OIF, the AFFOR Rear group understood that its job was not to support ACC/CC but to support the COMAFFOR as represented by the CENTAF A-staff. The distinction is very important and was understood by all involved. As there could be conflicts of interest between the COMAFFOR and ACC/CC, they understood they worked for the COMAFFOR, not ACC.

In October/November 2001, ACC/CE advocated that an A-7 be established forward to handle installation development of the austere FOLs used during OEF. This position (the A-7) was established in February 2002. The rationale was that the ACC/LG could concentrate on weapons systems sustainment and support while the ACC/CE and CENTAF A-7 concentrated on installation beddown and construction. The AFFOR A-4/A-7 in the rear at Langley AFB remained a combined function, with senior civil engineers and logistics colonels alternating chairing of the senior position. For assistance resolving issues that were raised to CENTAF Rear at Langley, they reported to their respective civil engineering and logistics home organizations.

Although not corresponding to their doctrinal responsibilities, the AFFOR A-4/A-7 functions were performed in the Combined Air and Space Operations Center (CAOC). Doctrine calls for the separation of the A-staff functions and the CAOC functions. Although not technically a break in doctrine, duties and responsibilities could become confused in collocating A-staff and CAOC personnel. Traditionally, and according to doctrine, the CAOC is responsible for developing the Air Tasking Order (ATO). Consequently, the logistics contingent in the CAOC was responsible for assessing resources needed to support the ATO and assessing effects of resource shortages. The AFFOR A-4 staff, on the other hand, needed to concentrate on assessing support effectiveness of alternative deployment and employment concepts identifying constraints to the A-3/-5 staff and the Air Operations Center (AOC), although it does not work for the AOC/CAOC.

These roles are similar to those indicated in the CSC2 TO-BE operational architecture, which states that the AFFOR A-4 and A-6 Forward would perform, plan, and assess support needed to meet the needs of the air campaign. The A-3/-5 would establish many of the requirements for support then the CAOC would develop the ATOs to execute the plan. The A-4 and A-6 would work with the A-3/-5 to develop the campaign plan and assess operational plans to determine feasibility and resource implications of alternatives (Leftwich et al., 2002), including beddown assessments. During OEF, the AFFOR A-4, A-7, and A-6 staffs spent the largest proportion of their time dealing with beddown issues.[5] Because of the rapid response needed to conduct beddown assessments, the limited access to some sites, and the advancement of reachback technology to conduct initial site-feasibility assessments remotely, CENTAF A-4 Rear conducted many initial feasibility assessments.

The placement of the AFFOR A-staff is important. It should allow the AFFOR A-staff to concentrate on A-staff responsibilities of the system-wide combat support planning and execution role of the A-staff function—and not encourage too much attention on CAOC daily ATO production tasks. The A-4 in OEF indicated that colloca-tion provided easy access to the JFACC/COMAFFOR and the AOC for coordination. The A-4 also indicated that A-4 functions were kept distinct during OEF—a situation that may be difficult to achieve in future operations under different leadership.

The A-3/-5 also defined roles on an ad hoc basis. While the A-4/A-7 Rear was established at ACC, the A-3/-5 reached back to Shaw AFB for support—an arrangement that made it difficult to de-velop an integrated campaign plan. Problems arose in developing TPFDD inputs, because the JOPES input capability was in the A-3/-5 function at Shaw, but the A-4/A-7 inputs came from the AOR and Langley. The A-3/-5 function moved from Shaw to Langley in Octo-ber to alleviate coordination problems. However, in November, the A-3/-5 Rear function moved back to Shaw.

Furthermore, CENTAF had to support both Operation South-ern Watch and OEF from a limited number of sites. It had to con-duct beddown surveys and configure the combat support network in

---

[5] Interview with CENTAF/A-4 LGX staff, September 2002.

the midst of supporting permanent Operation Southern Watch rotations. While CENTAF provided deploying OEF forces with forward beddown support, some operational units associated with Air Force Special Operations Command (AFSOC) and AMC developed reachback capabilities to their parent commands for sustainment support, bypassing CENTAF for reachback assistance for some items but relying on CENTAF for sustainment support for other items such as FOL support. At times, this splitting-bypassing practice led to confusion about which command had responsibility for support. There was no doctrine to guide these activities, and reachback support was developed on an ad hoc basis.

As in JTF NA, staff augmentations were developed on an ad hoc basis, with no clear delineation of which functions were to be performed forward and which were to be performed in the rear. However, based on experiences in JTF NA, organizational roles and responsibilities evolved more quickly.

One issue affecting CSC2 development in OEF was the speed with which the operation was executed. Operational planning took place very rapidly. The President issued his letter to Congress on the American Campaign Against Terrorism on September 24, 2001. The combatant commander issued an operational order on September 25, 2001, and operations commenced on October 7, 2001. Figure 3.4 shows the OEF operational planning timeline and how CSC2 command lines evolved over time, including some forward and rear COMAFFOR A-4 functions. The A-4 functions were not guided by doctrine or published guidance that identified the specific processes and functions that each CSC2 node would perform.

CSC2 Nodes and Responsibilities in OIF

Many of the issues identified after JTF NA and OEF were not repeated in OIF. The long time in which to plan and define relationships, coupled with the Air Force's agreement on and initial implementation of a CSC2 operational architecture, greatly enhanced the command and control of combat support during OIF.

Figure 3.4
OEF Operational and CSC2 Timeline

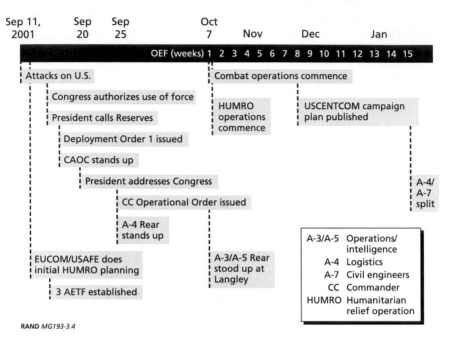

Before OIF started, AF/IL and the combat support community began implementing the designs from the TO-BE CSC2 operational architecture. Doctrine was in review, and some changes had been incorporated in doctrine. Roles and responsibilities were tied to specific organizations. When an organization did not exist but was required, it was created.

The Air Force was well prepared to conduct operations in Iraq. Many standing organizations used during OEF were still in place. The leadership had recent combat experience; most leaders were in place for at least part of OEF, and many had also held key positions during JTF NA. Individuals and organizations were better prepared to meet their responsibilities. The command structure was well defined. CENTAF acted as the supported command, and the rest of the Air Force supported CENTAF. The COMAFFOR had an understanding of expeditionary operations. The A-4, A-6, and A-7 all had experience in the AOR and were familiar with their responsibilities. At the Air Staff, the AF/IL had been in place prior to the events of September 11, 2001. He had built teams, defined roles, and fostered an attitude of continuous improvement. Both at the division level and in the Combat Support Center, part of the Air Force Crisis Action Team (CAT), his staff was experienced and well trained.

Some of the functions performed in the CAT had been in place since September 2001. Most MAJCOMs stood up a battle staff to support the OIF effort. The ACC/CC was quick to define roles to his staff: His entire command would support the Combined Forces Air Component Commander (CFACC) and the CFACC's staff. The ACC/LG, the ACC/CE, and the ACC/SC all worked together to ensure warfighters had the required support. The AFMC staff had been at or near surge operations for OEF, and many of their wartime operations were still in place.

The success of this operation was aided by the expeditionary mind-set of its leaders. For example, AF/IL started moving materiel forward to the AOR before the official requirement, thus increasing the speed of deployment. The success of the combat support portion of this operation was due in part to the Chief of Staff of the Air Force. His foresight and guidance enabled the Air Force to be prepared to conduct expeditionary operations. His "Chief's Sight Pictures"[6] and regular exchanges of information enhanced the Air Force's ability to truly accept and institutionalize the AEF concept. Phrases such as "accept deployment as the norm" and "adapt to the battle rhythm" both enabled this operation and motivated the forces.

Previous experiences have shown that poorly articulated goals caused wasted resources and effort. We found no evidence of such a lack during OIF. The COMAFFOR and all who supported him understood the operational effects that could be enabled or enhanced by a given combat support requirement. For example, there was a requirement to increase the size of the concrete parking ramp at a designated base in the AOR. The requirement was defined as a larger

---

[6] "Chief Sight Pictures" provide senior leader viewpoints on current topics. They can be found on the Air Force website at www.af.mil/viewpoints.

ramp that would allow additional B-1 bombers to be parked. Those bombers could deliver significantly more JDAMs per day to support combat operations. Another example is the refueling option implemented to alleviate a fixed storage limitation at another location. The refueling option enabled four times more reconnaissance sorties per day.

Planning for an operation against Iraq had been ongoing since shortly after the end of Operation Desert Storm in 1991; however, specific operational planning began as early as March 2002. When the President of the United States declared success in OEF, many planners and senior Air Force leaders shifted their attention to Iraq. In this report, March2002 is used as the beginning point because that was when planners focused their attention on potential operations in Iraq.

Figure 3.5 illustrates how CSC2 command lines evolved before and during OIF. In July 2002, Central Command planners began to define requirements and created an initial operational plan. A Warfighter Support Conference, hosted by AFMC and facilitated by AF/IL, was conducted in August 2002. The conference was used to address essential elements of the Logistics Sustainability Analysis (LSA)[7] that was used to help build the combat support plan for OIF by allowing numerous participants from various MAJCOMs and functional areas to review the initial operational plan and identify actions that needed to be taken to support the plan.

Following the conference, the CENTAF A-staff began working with the COMAFFOR to define alternative courses of action for when combat support shortfalls would impede the operational plan. Numerous actions were taken to move munitions, vehicles, bare base assets, and fuels support equipment forward. Combat support moved materiel forward early in a manner never before achieved.

---

[7] A Logistics Sustainability Analysis is an analysis of agile combat support shortfalls and limiting factors of an operational plan. The final report, called the LSA, is presented to the J-4.

Figure 3.5
OIF Operational and CSC2 Timeline

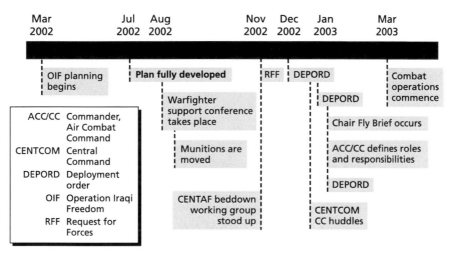

In January 2003, the COMAFFOR hosted a meeting called the "Chair Fly." By having all leaders together in one place, this meeting allowed the COMAFFOR to hold discussions with his Wing Commanders and the entire A-staff from both CENTAF and ACC. His goals and expectations were clearly defined. Each member of the A-staff was given the opportunity to outline problems and issues associated with achieving the desired results. Attention was focused on areas that would provide the greatest overall return on investment of time or resources.

Once the operational plan was agreed upon, the Air Force developed a time-phased force flow to execute the plan. Unfortunately, at the time of execution, the combatant commander changed the flow of forces, presumably to accommodate operational requirements. The Air Force had spent extensive time developing a plan and placing just the right amount of support with the respective warfighters. When the Deployment Orders (DEPORDs) did not match the developed force flow, the Air Force encountered problems adapting to the changes. The extensive coordination required when initiating changes in JOPES was a major area of concern.

JOPES can adapt to changes. However, the Air Force and the joint community have adopted many rules and processes that hamper its ability to respond *quickly*. JOPES is an excellent tool for deliberate planning; however, to add or subtract something from the TPFDD requires that several individuals approve the change. These approvals are required to ensure that everything on the TPFDD flows at the correct time. During crisis action execution, time is not always available to complete the same approval process. JOPES is not flexible enough and was not designed to be tailored at the execution level.

In addition, the Air Force has a limited number of personnel trained in JOPES. For example, during any given shift of the ACC CAT, there may be only one or two people qualified on JOPES.

Several other factors also contributed to the problems the Air Force faced with the TPFDD during OIF. Among them was the level of detail the Air Force places in the TPFDD. On the one hand, the U.S. Army has one unit type code (UTC) for an entire light infantry division.[8] The Air Force, on the other hand, has one UTC for a flying squadron and several associated UTCs that are necessary to enable that flying squadron to have warfighting capability. The Air Force has several "one-each"-type UTCs, many of them for Expeditionary Combat Support (ECS) vehicles. Everything in the Air Force is a trade-off and thus requires more detail and flexibility in the TPFDD than in the Army.

During the planning for OIF, the CENTAF A-4, A-6, and A-7 recognized that, once forward, they would need reachback support from ACC. Reachback to supporting MAJCOMs has been enacted in supporting several of the recent military operations. However, in each case, support was developed at the time of execution. As discussed earlier in this chapter, there were no clearly defined or documented

---

[8] Interestingly, the RAND Arroyo Center has evaluated the Army single-UTC concept and is recommending that the Army use the same process as the Air Force in constructing more modular packets of forces for a more expeditionary deployment. In addition, an Army infantry division may have one UTC, but there are many associated UTCs for nondivision assets to support the division and the theater.

functions for the A-4 Rear to accomplish. Each contingency incorporated differing approaches.

During OIF, roles and responsibilities were established early and organizations were given the authority necessary to perform their assigned responsibility. For example, the CENTAF A-4 Forward staff concentrated on day-to-day operations. The A-4 Rear staff, remaining at Shaw AFB, worked on beddown issues for new FOLs. The A-4 functions at ACC, in the ACC/LG, and the A-4 operations officer in the ACC CAT, worked to support both CENTAF Forward and CENTAF Rear. In the Pentagon, the Air Force Headquarters Air Staff A-4 functions, the Combat Support Center, and the Agile Combat Expeditionary Support (ACES) team monitored the situation and worked global issues. In addition, the Beddown Working Group was initially established at CENTAF Rear to facilitate combat support issues. However, once the group began to meet, all functional areas attended the meetings and numerous functional issues were resolved. The command relationships established to support OIF were consistent with the CSC2 TO-BE operational architecture (see Figure 3.6).

However, the A-3/-5 did not define roles and responsibilities or establish reachback procedures in the same manner as did the A-4, A-6, and A-7. As it did in OEF, this split-A-staff reachback arrangement led to problems. One problem was not having the right individuals together to resolve issues immediately. Coordination of roles and responsibilities took time to develop.

In OIF, the COMAFFOR was also the Combined Forces Air Component Commander and, as such, oversaw the CAOC and commanded the U.S. Air Force forces in the AOR. While the Air Force could easily support this arrangement, the CFACC could have been a Navy admiral afloat, causing a split in the command and control processes. In OIF, the A-4 Forward assisted the COMAFFOR with specific requirements. The lines of authority and responsibility were established during OEF, and lessons learned from previous engagements were not repeated. ACC stood up its own A-staff to sup-

Figure 3.6
CSC2 Organizational Structure During OIF

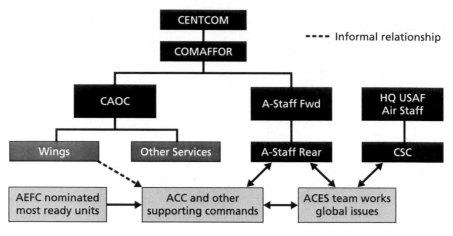

CSC   Contingency Support Center
AEFC   Aerospace Expeditionary Force Center
ACC   Air Combat Command
ACES   Agile Combat Expeditionary Support
RAND *MG193-3.6*

port OIF and the COMAFFOR. Functional representatives from ACC assisted their counterpart functional representatives at CENTAF.

Comparison of Actual and TO-BE CSC2 Nodal Responsibilities

At the Air Force level, the TO-BE CSC2 operational architecture calls for the Contingency Support Center (CSC) to monitor combat support requests for a particular contingency and assess the impacts of those requests on the Air Force's ability to support the current and other potential contingencies. During OEF, the existing Combat Support Center acted as the CSC and intervened when necessary to allocate scarce resources to the AOR. The Combat Support Center did some real-time quick assessments for FOL support assets. However, the Combat Support Center relied on the supporting MAJCOMs to supply weapons system supportability assessments and provide the effect of OEF operations on peacetime training and other potential contingency operations. The operational architecture calls for a CSC to conduct these weapons systems and FOL support assessments.

Part of the Combat Support Center, the ACES team performed the functions of the CSC for OIF. In addition to conducting integrated assessments, such as base support, the ACES team tracked and monitored 117 action items identified in the Logistics Sustainability Analysis and worked to find solutions for competing demands for scarce resources. The ACES team took a global view of combat support, which meant that, while working for the AF/IL, it was able to cross MAJCOMs and theaters to find optimal solutions. This same team (ACES) may be able to support quantitative assessments needed to support the Program Objective Memorandum (POM) process during non-contingency operations (Hillestad et al., 2003).

Similarly, AFMC/LG assumed many of the responsibilities identified with a future Spares Inventory Control Point (ICP) in the CSC2 architecture, such as tracking shipments and working more closely with customers and suppliers. Established at Headquarters, AFMC, in 1994, the Logistics Support Office has an analysis section that monitors shipment pipelines in order to correct backorder problems, depot processing issues, and contract problems. The Logistics Support Office also tracks the delivery times to various locations by various commercial and military modes of transportation. It coordinates with Air Mobility Command, commercial carriers, and personnel in the AOR to alleviate shipping problems. If a particular shipment was delayed, the Logistics Support Office's CONUS Distribution Management Cell was empowered to reroute shipments. Delivery-time information was relayed to customers so that they would be able to make better decisions about transportation modes for future shipments. The Logistics Support Office also shared specific shipping times with the respective commercial shippers and was able to get those shippers to modify their behavior and improve by either adding or modifying routes. Many lessons were learned during OEF. By the time OIF commenced, the Logistics Support Office was fully able to leverage commercial capability ahead of the requirements.

During OIF, AFMC, along with its customer MAJCOMs, continued to use the High Impact Target (HIT) list that was developed during OEF. The HIT list consists of each MAJCOM's most important repair parts, as identified by the MAJCOM. AFMC monitors the parts in the various Air Logistics Centers. It has automated many of the processes associated with maintaining the list and gathering status reports.

Learning from past experiences, AFMC established the Warfighter Sustainment Division to specifically address problems with wartime combat support. The Warfighter Sustainment Division was created to be a single interface between AFMC and its customers. The two sections of this division are the Operations Branch and the Logistics Analysis Branch. The Operations Branch concentrated on tracking shipments, coordinating repairs through the system, and responding to customer queries. The Logistics Analysis Branch provided forecasting and attempted to identify shortfalls and issues.

Although not formally implemented, pieces of the CSC2 TO-BE operational architecture were already in existence before OIF began. Both AFMC and the ACES team have taken steps to assume responsibility for specific nodes of CSC2 processes.

Integrated Closed-Loop Assessment and Feedback Capabilities

Another of the key concepts established in the CSC2 TO-BE operational architecture is a closed-loop assessment and feedback process. This process can inform operations planners of the impacts of combat support process performance on operational capability. In operational planning, this concept has been well understood and has been the topic of doctrine for a long time (Boyd, 1987). For example, in operations planning, it is standard procedure to conduct battle damage assessments. If the assessments reveal that some targets have not been destroyed or rendered unusable, the planners modify the ATO to retarget.

In both JTF NA and OEF, not very much attention was given to combat support closed-loop feedback processes or to relating combat support process performance to operational goals. In fact, many of the feedback and performance measures for combat support processes were incomplete. Effective use of information feedback in combat support planning and control is dependent on two things: (1) reliable access to information and (2) a framework for measuring combat support process performance against goals or standards that are needed to achieve operational goals in the specific contingency operation.

Many support decisions, such as those for capacity, manpower, and thresholds, were made without knowledge of how those decisions may affect operational goals. Most combat support processes lacked the data-tracking capability to tie their actual levels of performance to those that were planned for achieving specific levels of operational capability. In fact, many support decisions were not based on operational needs or system status. Combat support response time goals were set arbitrarily or were based on historical performance, not on operational requirements.

Improvement was evident during OIF. Lessons learned from previous operations and the amount of planning and preparation for OIF accounted for part of this improvement. Planning and preparation began with the Warfighter Support Conference, which was held in August 2002, seven months before operations began. CENTAF planners worked with MAJCOMs and Air Force–level logistics experts to review requirements and establish a plan to accommodate shortfalls. The ACES team at the Pentagon tracked and monitored the execution of action items. After the conference, all MAJCOMs were in agreement, producing a concentrated effort to support the operational plan.

At the same time, several key command and control decisions were made. The Air Force was willing to take a risk when it decided to move munitions and other heavy logistics assets to the theater ahead of an approved execution order. Having most of the required materiel in the AOR was preferable to having the materiel waiting in CONUS.

While the Warfighter Support Conference offered the opportunity for many to help determine how to support the operational plan, it also gave insight into what the COMAFFOR hoped to accomplish. This understanding led to agreement on what should be reported and on a standardized reporting format. Web-enabled tools helped produce a picture that could be viewed at various levels of command, allowing an understanding of the requirement throughout the logistics arena. Customers and providers could view and download information immediately. During OEF, this same information exchange took numerous telephone calls and e-mail messages to achieve.

Since OEF, progress has been made on developing information that could be used in combat support execution planning and control. For instance, several AFMC initiatives, the Strategic Distribution Management Initiative (SDMI), Total Asset Visibility, and other system improvements have developed information feeds to track current values of various transportation pipelines and other combat support process performances.

Not as much progress has been made on developing a combat support closed-loop control framework. Support information that was tracked was not always used in decisionmaking. Information collected and analyzed by the AFMC Logistics Support Office was transmitted to customers, but that information was not always acted on. SDMI tracked current, or "AS-IS," performance, but it did not base performance goals on operational needs. The essential elements of a combat support closed-loop control framework and a more detailed description can be found in the CSC2 TO-BE operational architecture (Leftwich et al., 2002).

Implications

The Air Force had the advantage of OIF commencing as Air Force operations in Afghanistan were drawing down. Many of the leaders were already in place. Organizations built on an ad hoc basis for OEF were refined for OIF. Early in OIF planning, roles and responsibilities were clearly defined and articulated. Given adequate time, a solid plan was developed.

The Air Force should ensure that the lessons, both good and bad, from this operation are passed on to future leaders, perhaps through doctrinal changes. Doctrine should institutionalize success from past operations. An expeditionary Air Force (regardless of how it is organized and managed) will be required in the future. Training and equipping leaders to deal with expeditionary operations will continue to be a challenge. Much should be learned from the way roles and responsibilities were defined during OIF. Organizations used in recent operations should become standing organizations, exercised regularly.

Combat support planning needs to be integrated in the operational campaign planning process. The effects of alternative combat support strategies, tactics, and configurations need to be known to operations personnel when a plan is selected. Once a jointly developed operations and combat support plan has been determined to be feasible (through an LSA) and capable of achieving the desired operational effects, a closed-loop feedback and control system needs to track actual combat support process performance against planned values. When the system exceeds control parameter limits, the CSC2 system needs to signal combat support personnel that corrective action is needed. The CSC2 operational architecture outlines how this planning and control could take place across the echelons of support and throughout the phases of operational campaigns (Leftwich et al., 2002).

The CSC2 operational architecture also specifies CSC2 nodes and associated responsibilities that are consistent with those that were developed and used during OIF. The CSC2 architecture specifies the broad responsibilities of the COMAFFOR A-Staff Forward and Rear, the Contingency Support Center, and Inventory Control Points. Many of the COMAFFOR A-Staff functions, as well as those of the other nodes, can be performed by standing organizations.

Command and control reachback support needs to be defined for all A-staff functions. Should reachback be separated by functional responsibility or should all A-staff functions be collocated in standing rear organizations that can serve more than one COMAFFOR? Collocated reachback is the recommendation presented in the CSC2 TO-BE operational architecture. For instance, if another NAF were to have a sizable operation at the same time that 9th Air Force was engaged in OIF, reachback for the new operation could be conducted from ACC in a section of an Operations Support Center that could support that operation while another section continued support for OIF.

To implement the CSC2 architectural concepts requires changes to doctrine, education and training, organizations, and systems. AF/IL has initiated doctrinal changes to begin implementation of the CSC2 processes and standing organizations—a step in the right direction. Much more is needed. Education and training programs are needed to teach these concepts. The expeditionary mind-set should be incorporated in all doctrine, policy, education, and training. Both current leaders and future leaders should be prepared for expeditionary operations.

In addition, decision support systems are needed to carry control information to combat support personnel so that significant deviations from planned performance can be corrected before operational effects are felt. The use of information systems has improved, but additional capability is needed. For example, better access to control information needs to be provided through automated system interfaces. The wider use of automated tools would enhance beddown assessments. Better links are needed between operational requirements and AFMC process performance and resource levels.

The systems that support TPFDD development in the deliberate planning cycle should also be able to support crisis action execution.[9] One possible solution may be for the Air Force to offer a single UTC to the combatant commander, as the Army does, and then to internally tailor that UTC as required to meet the mission requirement. This concept was applied already in the communications area

---

[9] Chairman of the Joint Chiefs of Staff Memorandum 3122.02C is in draft. This memorandum will provide guidance for the exercise of authority by combatant commanders in operations using JOPES.

during OEF and OIF. The Theater Deployable Communication UTC (6KTDC) is tailored into smaller taskable UTCs of 6KTEA, 6KTEB, 6KTEC, etc. Another option may be Force Modules, which the Air Force leadership is currently developing. A Force Module is a prearranged force that would be able to provide a given capability. Yet another option is a new system whereby inputs are the number of fighters or bombers to be deployed and the output is a list of UTCs. Or perhaps the input could be a capability (the ability to provide close air support for a given size army force for X days, and deep strike capability to destroy Y number of targets and all associated ECS to bed down and sustain forces for Z days). Not only would the tool add UTCs, it would look for redundancy and define the requirements.

CHAPTER FOUR

Forward Operating Locations and Site Preparation

Selection and development of FOLs play an important role in meeting AEF goals. In this chapter, we discuss timelines for forward operating locations and combat support efforts associated with preparing for deployed forces. We compare the three recent operations, JTF NA, OEF, and OIF, and present implications for future operations and the AEF concept.

Findings

Our findings are in the following areas:

- Forward operating location timelines
- Host-nation support and site surveys
- FOL development and construction
- Contractor support.

JTF NA FOL Timelines

The development of FOLs in JTF NA varied greatly, as did their development in OEF and OIF. Figure 4.1 illustrates the time involved to develop FOLs for JTF NA relative to a notional AEF goal, to deploy a force package to a familiar or known FOL and begin opera-

Figure 4.1
FOL Development in JTF NA

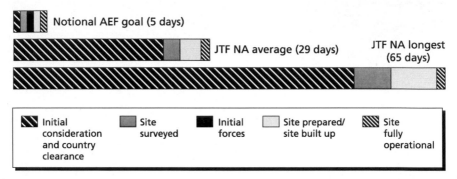

tions within a few days (the top timeline in the figure). For example, the current goal allows 72 hours for deployment, with initial operations beginning within 48 hours after arrival, a 5-day notional goal. The other two timelines show the average time to develop an FOL during JTF NA (29 days) and the longest time to develop an FOL during JTF NA (65 days).[1]

Following the legend for the FOL site-development timeline, located under the bottom timeline,[2] we see that the first section of each timeline represents the time it took to get country clearances once the site has been considered as a potential FOL. The second section represents the time it took to do an initial site survey. The next section shows the time it took for site preparation and site development. The end section of each timeline shows when the base was considered fully developed, or fully operationally capable. We define a *fully operationally capable* (FOC) site as a site with a full complement of base

---

[1] Information is based on extensive interviews and data collected from USAFE and CONUS-based Air Force organizations.

[2] The JTF NA timelines do not include initial force arrival; the OEF timeline does.

operating support in all functional areas, not necessarily a site that has closed all its TPFDD requirements.[3]

Note that during JTF NA, the largest portion of time was spent obtaining country clearance and handling the diplomatic issues to bed down forces at specific sites. Note also in comparing the total time needed to develop sites that the 5-day notional goal was greatly exceeded.

OEF FOL Timelines

As in JTF NA, some FOL locations used in OEF were adequately equipped to host U.S. forces with little buildup and other sites were bare bases and required significant development. Diego Garcia was fully developed quickly, in approximately 17 days, whereas Jacobabad took approximately 78 days to become fully capable to support operations.[4]

Figure 4.2 illustrates the development of several FOLs during OEF. Note that these timelines start when each site was initially considered for deployment, not from one specific starting date: Site development began at different times for each site. The third section on each timeline shows the arrival of the initial airmen at each site. The airplane symbol shows when each base received aircraft and began conducting missions, or was initially operationally capable (IOC). As shown in the figure, operations were conducted from FOLs well before they were fully developed.

Diego Garcia became a fully functioning FOL very quickly (relative to other sites) during OEF. Prior to OEF, the Air Force had determined that Diego Garcia would be a bomber island—that is, one of several predetermined, prepared, forward operating locations for heavy bombers. As a result, the Air Force has had extensive

---

[3] *Fully operational* does not mean that all materiel and personnel needed at a site are present. It refers to resources that are in place or developed to meet mission needs. The definition of an FOC site used here is roughly consistent with green lights for all functional areas of the stoplight charts used in the Headquarters AF/IL External Slides from the daily briefings.

[4] Data are from Headquarters AF/IL External Slides from the daily briefings.

Figure 4.2
FOL Development in OEF

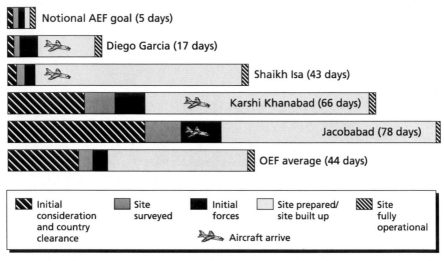

experience in deploying and operating out of Diego Garcia, and some bomber island preparations were already under way. Thus, the FOL preparation time was less than that of other FOLs that were not as developed or whose capabilities were unknown to Air Force planners.

Jacobabad and Karshi Khanabad, in contrast, required extensive buildup of water-purification, sanitation, and power-generating capabilities that had not been required at pre-planned FOLs such as Diego Garcia. During OEF, the largest portion of time was spent preparing the sites. It took time to set up force protection, repair deteriorating parking ramps, set up communications, build munitions pads, and develop large tent cities. The FOL development timelines were much longer for these unanticipated FOLs.

OIF FOL Timelines

As in JTF NA and OEF, the development of FOLs in support of OIF varied greatly. Some FOL locations were adequately equipped to host U.S. forces with little buildup; other sites were bare bases and required significant development. Figure 4.3 shows a sampling of the development of bases involved in OIF. Again, the timeline for developing each FOL begins when the site was initially considered for deployment. Site development began at different times for each site.

As in Figures 4.1 and 4.2, the top line illustrates the AEF goal, a 5-day notional goal. The next two lines, Karshi Khanabad (K2) and Incirlik, illustrate the time it took to prepare sites at which the Air Force already had forces in place. Karshi Khanabad was an unanticipated, austere FOL requiring extensive development during OEF.

Figure 4.3
FOL Development Timelines Varied in OIF

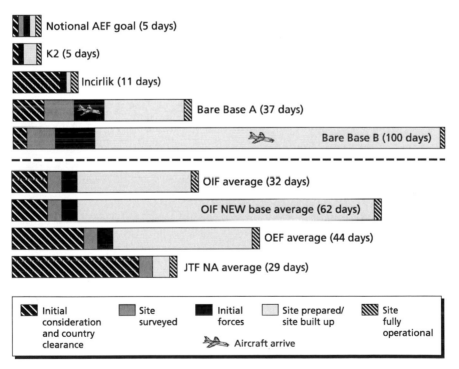

The development of Karshi Khanabad was completed prior to OIF, so very little was required to prepare the site for use during OIF. Incirlik AB, Turkey, on the other hand, required more time to develop. Although Air Force forces had been providing support for Operation Northern Watch out of Incirlik since the end of Operation Desert Storm, Turkey denied the Air Force use of Incirlik for supporting OIF. These timelines show that even at existing bases, FOL development required more time than the AEF goal.

The next two timelines in Figure 4.3 show the development of two bare bases during OIF, which for political sensitivities are identified as Bare Base A and Bare Base B. Although both bases required a long time to develop, it is important to remember that there was adequate time to plan for OIF. These timelines represent the time used in preparation for OIF and do not indicate how quickly these actions could have been completed had the requirement been different.

The bottom four timelines provide information about the average amount of time needed to develop FOLs during the three recent operations.[5] The average in the first timeline, for FOL development in OIF (32 days), includes approximately 14 bases that were developed in support of OEF and then used again to support OIF. Their development for use in OIF took only a few days at the most; so the total average is less. The second timeline shows the OIF average if the 14 OEF bases were not included in the total average. The average becomes considerably longer, taking 62 days to develop the bare base FOLs used in support of OIF.

In comparing JTF NA, OEF, and OIF, note that during JTF NA the largest portion of time was spent getting country clearance and handling the diplomatic issues to bed down forces at specific sites. During OEF and OIF, the largest portion of time was spent preparing the sites. More time was spent on country clearance during OEF than during OIF.[6]

---

[5] The JTF NA timeline does not include initial force arrival; OEF and OIF timelines do.

[6] Many other factors affected FOL timelines, and especially the development of FOLs, including transportation shortfalls, availability of resources, the Guard/Reserve mobilization process, and transportation of engineers from CONUS.

JTF NA Host-Nation Support, Country Clearances, and Site Surveys

During all three operations, host-nation support agreements had an important effect on FOL site development timelines.

During JTF NA, host-nation support was slowed by both policy and political barriers. Policies for obtaining host-nation support and country clearance for conducting site surveys were not clearly delineated. Whether the operational community or the logistics community would request the necessary host-nation support to conduct the site surveys was unclear.

In addition, no standardized site-survey checklist had been developed. The base support planning policy identifies a list of areas that should be addressed by the survey team, but that list is designed for teams that have time to conduct a lengthy survey. During JTF NA, host nations were slow to grant country clearance: It took approximately 12 days to receive the clearance. Once clearances were obtained, teams had a very limited time in-country, often only one day. The detailed, deliberate planning checklists were not suited for the type of survey conducted during JTF NA.

OEF Host-Nation Support, Country Clearances, and Site Surveys

With limited site-survey information available for some sites, CENTAF planners often had to rely on promised host-nation support rather than on detailed site surveys to accomplish initial site planning during OEF. Host-nation support and country clearance permissions were often delayed. In some cases, host-nation support was promised but not delivered, or it was slow to evolve.[7] The same issues encountered during JTF NA were faced again during OEF.

Air Force personnel routinely worked with U.S. Embassy staffs on such host-nation issues as diplomatic clearances. Beddown sites were located in ten different countries during OEF, and each site requiring host-nation negotiations. These staffs of the embassies are equipped to do their peacetime job, but are not adequate for wartime operations.

---

[7] Interview with CENTAF/A-4 LGX staff, September 2002.

Obtaining clearance to enter a country was often difficult, but obtaining access to specific sites was often more difficult. Specific access must be granted for site-survey teams to enter a potential FOL site. Often, survey teams were granted country access but not site access, which caused delays.[8] In some cases, access to potential beddown sites was denied.

Host-nation support changed as operations unfolded during OEF. For example, in late October/early November, Qatar, an important host nation, closed Camp Snoopy, located at Doha International Airport. Units were forced to relocate to Al Udeid Air Base outside of Doha. Resources were consumed as units moved from an FOL under development to another site. These moves were outside the control of the Air Force and caused delays.

OEF required establishing FOLs in areas that had not before been considered as potential beddown locations. Services were caught with little or no site data for operations in the Afghanistan AOR. By itself, the use of unanticipated sites would have been a challenge. However, the challenge was further exacerbated by the lack of a standardized site-survey process among Air Force commands, U.S. military services, and allies; a problem that was also encountered during JTF NA.

The lack of survey standards and a common site-survey tool complicated the survey process. Initial CENTAF and Central Command site surveys were conducted on an ad hoc basis. Survey assessments were done by multiple agencies, and the assessment procedures were not uniform. Coalition teams, which did not always include civil engineers (Barthold, 2002, p. 2), were sent out with hard copies of a checklist without receiving training on how to conduct a survey using the checklist. Consequently, the checklists were returned incomplete, many in different forms, adding delays to deployment timelines.

Politics also played a role in site surveys. Team composition was influenced by political sensitivities, which led to additional problems

---

[8] Interview with CENTAF/A-4 LGX staff, September 2002.

and delays. Air Force and coalition partners did not share common standards and expectations on the contents of site surveys. When coalition partners were in charge of conducting surveys with Air Force support, very different products were produced from those for which Air Force personnel led the process. Even Air Force–led surveys were nonstandard.

Within the Air Force and among the services, site survey tools and techniques differed by command and mission type.[9] Most combat support assessments were done quickly and manually, and they were not uniform in quality. Existing USTRANSCOM and AMC information on the AOR was not rapidly shared with Central Command and CENTAF.

Another challenge was the lack of a global site-survey database for information gathered during site surveys. Existing data are stored in many places where access to data is controlled by the owning command. Often, information may not be shared from one command to another or from one service to another. Information gathered during site surveys should be stored in a common database for use by all services.

Two site-survey tools are currently available to the services for use: Survey Tool for Employment Planning (STEP) and Beddown Capability Assessment Tool (BCAT). STEP/BCAT standardizes the data collection approach and uses computer-generated templates to complete survey information. These systems were not used to collect data. Site-survey personnel were not familiar with STEP/BCAT; they did not wish to travel with classified equipment; or they lacked the equipment and communication lines to update the Knowledge Database located at Maxwell AFB (Günter Annex).

OIF Host–Nation Support, Country Clearances, and Site Surveys

Host-nation support and country clearance permissions were also delayed during OIF, lengthening the time needed to develop FOLs. In some cases, host-nation support was promised but not delivered. In

---

[9] The Air Force is currently in the process of developing an Integrated Site Survey Team Checklist to standardize the survey process.

other cases, access to potential beddown sites was denied or delayed, as during OEF. Even when host nations agreed to allow forces to use their facilities, they often asked that their support not become public knowledge. Seventeen different countries were used to bed down forces during OIF. Each country required host-nation support.

As in OEF, host-nation support changed as operations unfolded. In other instances, support the planners felt assured they would receive did not materialize. For example, expected support from Turkey did not exist.[10] Prior to OIF, the Air Force was operating out of Turkey in support of Operation Northern Watch. Most Air Force planners expected that because Turkey was a NATO partner and long-time friend of the United States, and because the Air Force was currently operating from Turkish bases, the Turkish government would cooperate and allow the United States to use Turkey's facilities during OIF. When the Turkish Parliament voted to not allow the United States to launch strikes from their soil, many were caught off guard and unprepared for the refusal. Not being able to attack from the north changed the battlespace and forced the Air Force to use tankers and conduct longer missions, creating additional support burdens.

During OIF, Air Force personnel routinely worked with U.S. Embassy personnel on such host-nation issues as diplomatic clearances, as they did during OEF. Learning from experiences during OEF, the Air Force established a special group to interact with embassy personnel. In some countries, a General Officer was assigned to the embassy to handle beddown and host-nation support issues. Reservists also augmented embassy staffs to assist with beddown issues.

Likewise, Air Mobility Command's Global Assessment Teams (GATs) were employed during OIF to survey and assist in the establishment of all the major airfields in Iraq. Their use significantly improved the site-survey process over that experienced in OEF.

---

[10] This is not intended as a political review of right or wrong, only a review of the effects on ACS of the decisions made by the Turkish government.

JTF NA FOL Development and Construction

Most JTF NA FOLs were well developed, as was the European Command theater infrastructure, including the transportation and supply infrastructures. Both commercial transportation options and local industry were available, so that the Air Force could use trucking, air, rail, and sea modes of transportation to meet deployment and re-supply needs. Moreover, the planning timelines enabled supply pipelines to be in place and operational as JTF NA began. During JTF NA, FOLs were developed in less austere locations than in OEF or OIF, so installation construction was not as extensive.

OEF FOL Development and Construction

In contrast, OEF depended upon opening numerous unfamiliar, unanticipated, and unprepared FOLs for SOF operations and for intelligence, surveillance, and reconnaissance operations in a very short time frame.

Some FOLs in the Central Command AOR were familiar to Air Force planners and had been used previously to support Operation Desert Storm, Operation Southern Watch, or allied exercises. Prior to OEF, the Air Force had personnel, aircraft, and equipment deployed at several bases in different countries in southwest Asia to conduct Operation Southern Watch, the enforcement of the no-fly zone over southern Iraq. Operations in Afghanistan required an augmentation of these forces (Burns, 2003, briefing slide 3). Some existing bases were supplemented with additional personnel and infrastructure. Bases that once housed prepositioned equipment became transportation hubs or beddown locations for tankers, ISR assets, or bombers. Bases close to Afghanistan housed tactical aircraft; bases farther away were used for bombers, ISR assets, tankers, C2, and airlift support. Although these familiar FOLs were developed rather quickly to support operations, they were a long distance from Afghanistan, requiring long flights and aerial refueling.

Other new bases were opened closer to the Afghanistan area of operations. FOLs in the immediate area of the conflict were not as familiar to Air Force planners prior to OEF and were not prepared for immediate use. Because of the austere locations of many of the

FOLs, extensive engineering and development efforts were required, as was the use of bare base support assets. Most host-nation facilities required improvements, because existing buildings and facilities were unusable. Many sites required extensive development. In several cases, the Air Force deployed to very austere locations and commenced operations before FOLs were fully developed. The Air Force accepted risks to deploy and employ the force quickly. Consequently, high-demand/low-density combat support personnel in, for example, combat communications, force protection, and civil engineering were required in significant numbers at most locations.

Eventually, bases within Afghanistan itself were also used during OEF (Burns, 2003, briefing slide 4). The short planning time made FOL development difficult. Poor infrastructure, including less-than-ideal roads and limited rail capability, further compounded the problem. In all, the number of personnel, aircraft, and beddown locations in the AOR to support OEF approximately tripled the Operation Southern Watch presence already in the AOR.[11]

The installation development performed by Prime BEEF during OEF was the largest development of FOLs since Vietnam. To complete FOL construction, 1,564 Civil Engineer (CE) personnel were deployed: 848 Active duty, 128 Reserve, and 588 Guard. Prime BEEF teams also conducted airfield support operations; fire protection; nuclear, biological, and chemical defense; explosive ordnance disposal; airfield damage repair; beddown; facility and utility system sustainment; and other functions. CE personnel were deployed to nine locations in October 2001. Construction projects included runway repair and ramp construction, as well as construction of facilities for the airmen.

In support of OEF, rapid engineer deployable heavy operational repair squadron engineer (RED HORSE) teams of 500+ people worked on 77 projects valued at approximately $70 million. In October, construction was planned or ongoing at seven sites: three Opera-

---

[11] See the CNN website, www.cnn.com/SPECIALS/2001/trade.center/military.map.html/

tion Southern Watch sites, three OEF sites, and one site awaiting construction.

In completing the construction work necessary during OEF, CE resources, both personnel and equipment, were stressed. For example, FOL support assets were available in Jacobabad and ready for construction; however, no civil engineers were available to assemble them.[12] FOL development timelines were delayed because of stressed CE resources.

OIF FOL Development and Construction

Buildup timelines for forward operating locations varied in OIF and depended heavily on preparation activities. FOLs that were partially developed or with which the Air Force had experience in previous deployments facilitated rapid force deployments. Some Operation Southern Watch and OEF bases were expanded and used in OIF. Other FOLs were developed at locations not used since Operation Desert Storm, and new bare bases had to be opened. Many austere FOLs were developed; however, with the long planning time, the use of these undeveloped FOLs was well planned and action was taken to allow them to be developed quickly. Some countries were sensitive about allowing the United States access and presence, which complicated beddown planning (Burns, 2003, briefing slide 5). Air Force planners had detailed knowledge of the AOR before OIF, which greatly enhanced their ability to open bases.

OIF required even more extensive FOL development than OEF. Several of the bare base FOLs used during OIF were located in austere locations where extensive engineering and development efforts were needed, as shown in Figure 4.3. Construction projects led 4,592 CE personnel to be deployed in support of OEF/OIF. The projects included runway repair, ramp construction, POL storage and distribution, as well as construction of facilities for the airmen. CE personnel worked on over 200 projects valued at approximately $329 million (Burns, 2003; U.S. Air Force, Central Command,

---

[12] Interview with Maj Gen Richard Mentemeyer, October 2002.

2003), as well as on new construction efforts, which stressed CE personnel and resources.

Several "firsts" occurred for CE personnel involved in OEF/OIF: the first integration of Army and Air Force engineers since World War II; the use of night-vision goggles for Rapid Runway Repair in Afghanistan; and employment of the Airborne RED HORSE. Airborne RED HORSE is a scaled-down RED HORSE capability that can be inserted onto a hostile or recently captured airfield to repair runways and enable other forces to follow.

Significant improvements also were made to the communications infrastructure during OIF. For the first time, communications was included in beddown planning. Communication Support built a communications architecture for the entire area of responsibility. This AOR-wide infrastructure included primary and backup communications paths, secured satellite communications, and reduced circuit activation times at deployed sites. Communication Support deployed 1,120 personnel and approximately 2,700 short tons of equipment in support of OIF.

Because of the long planning time, additional KU bandwidth was purchased for military use during OIF, a type of support not implemented during OEF because of its shortened planning timelines. Longer planning times allowed a more robust communications infrastructure to be developed. For example, the entire network had backup capability. At no time during the operation did any beddown location lose communications connectivity. Other advances in the communications area enabled five Communication Support personnel to stand up a voice network, both secure and unsecure (approximately one-half of a pallet of equipment), in about 5 hours at what was an enemy base just days before.

JTF NA Contractor Support

JTF NA relied on contractor support. The Air Force Contract Augmentation Program (AFCAP) provided support for U.S. forces by moving facilities, conducting paving and facility evaluations, and providing heavy equipment in places such as Bosnia, Hungary, Turkey, and Italy (Wolff, 2000). AFCAP also established a forward support location at Ramstein AB, Germany, from which it supported the construction of a small city of 17 kilometers of roads, 1,820 tents, 1,006 latrines, 270 water taps, 12 school areas, 44 bathhouses, and 176 food-preparation areas in 51 days (AFCAP Update, n.d.). Contractor support helped ease the installation development workload for the military personnel in JTF NA.

OEF Contractor Support

During OEF, contractor support already on-site in many locations aided the development of FOLs. Contractors at WRM storage locations were able to shift from maintaining WRM to preparing FOL sites at collocated FSL sites.

Contractors supported site preparation in five locations during OEF. They helped civil engineers establish Camp Snoopy, construct tent cities in two locations, set up fuel farms, and refuel aircraft until Air Force personnel arrived. They also operated power plants at several locations (DynCorp, n.d.). Contractors provided equipment, ground transportation, bottled water, furniture, facilities, cellular telephones, laundry services, and fuel. They worked 485,772 hours of overtime and catered 1,279,187 meals. Additional contractor personnel were hired and existing personnel were re-allocated to support the needs of OEF. See Table 4.1 for an example of the increased contractor workload during the first 100 days of OEF.

A contingency clause in the contractors' statement of work allows them to provide this type of additional support, and their support was extremely beneficial in meeting rapid FOL development and in aiding uninterrupted sustainment. However, one consequence of using contractors in this capacity was the reduced outload capability: The same personnel who are setting up a tent city cannot be loading aircraft with WRM at the same time. This issue will be discussed further in the next chapter.

OIF Contractor Support

During OIF, contractor support already on-site aided FOL development, just as in OEF. Contractor reception teams were on-site

Table 4.1
Contractor Support Surged During OEF

Type of Support	Average for OEF[a]	Pre-OEF Average
Direct mission support tasks	162	10
Tons of air cargo moved	1,554	400
Truckloads moved	705	200
Total tons moved	9,331	3,000

[a]Average calculated from data for October 2001 to January 2002.

at nine locations and helped set up Initial Housekeeping sets at three locations. In all, they moved 1,312 vehicles, 1,332 pieces of fuels mobility support equipment (FMSE), and 280 pieces of AGE, and line-hauled 7,753 truckloads (DynCorp, n.d.). Table 4.2 shows the amount of contractor support used during OIF compared with contractor support used during OEF.

However, not all contractor support proved successful. Because civilian contractors did not show up to do the contracted work, the Army experienced poor living conditions (Wood, 2003; Krugman, 2003). Cryogenic facilities that the Defense Contracting Management Agency had contracted for lacked proper quality surveillance over local suppliers, eliminating their use for military operations. Insurance premiums for contractors in a war zone were extremely costly, and a contractor could not be forced to go into a war zone. Both of these situations limited the availability of contracted support.

Table 4.2
Contractor Support Provided in OEF and OIF, for Comparison

Type of Support	OEF	OIF
Direct mission support tasks	825	707
Tons of air cargo moved	8,258	9,170
Tons of line-haul moved	95,831	109,479
Total tons moved	104,084	163,304

Implications

Selection and development of FOLs play an important role in meeting the AEF goal. As experienced in JTF NA, OEF, and OIF, many actions can be taken to decrease deployment times and reduce FOL preparation times.

As discussed in this chapter, large amounts of time were expended in all three recent military operations in gaining access to a country and to a specific FOL within that country. Even when FOL sites were known and anticipated, time was required to develop them. Without host-nation support and sufficient infrastructure already in place, the 5-day notional AEF goal is largely unobtainable. Perhaps the notional goal should be adjusted to a more attainable goal.

Engagement policies and programs to familiarize Air Force planners with facilities in countries that may be sites for future operations could potentially reduce country access time. Such programs as Partnership for Peace, in which knowledge of and improvements to FOLs can be gained through exercises and deployments by documenting the information in GEOReach[13] expeditionary site survey tools, should be encouraged. Knowledge gained through this and other programs that enhance military-to-military contact can help speed deployments to important areas around the world.

Where possible, a select number of future FOLs in likely sites should be surveyed for capabilities. Goals could be established in each AOR for surveying potential sites for future Air Force use. Funds could be put aside for accomplishing such surveys. In some cases, sites in potential conflict areas could be prepared in advance for rapid deployment. The Air Force is already using GEOReach expeditionary site survey techniques to survey potential FOL locations. These plans should be codified and accessible to all MAJCOMs through a web-enabled process.

Training some Air Force combat support officers in a fashion similar to Army Foreign Area Officers would produce country and

---

[13] GEOReach is a program that combines tabular data with a visual image to provide commanders with situational awareness.

area specialists.[14] These Foreign Area Specialists could augment embassies in the early stages of a conflict, when military staffs at embassies are often overwhelmed. These specialists could facilitate rapid country clearances, access, and host-nation support agreements. In addition, military staffs at embassies should be augmented during wartime.

Leveraging contractor capabilities to assist civil engineers in developing FOLs once a contingency begins is another method of decreasing FOL preparation time. As demonstrated in OIF, AFCAP[15] and other contractor capabilities, such as WRM maintenance contractors at FSLs, can be leveraged to aid civil engineers in sustaining FOLs. Although it may be desirable to have Air Force civil engineers complete the initial beddown planning and construction, capabilities to augment scarce Air Force personnel skills could be developed through such programs. Databases of contractor capabilities, similar to FOL site surveys, should be developed in areas in which potential conflicts may be likely.

---

[14] The Air Force does have a career-broadening duty similar to the Foreign Area Officer. There is also discussion in the Air Force about developing a more robust Professional Military Strategist program for language and cultural specialists.

[15] AFCAP is a contract tool, available only during contingency response, to provide civil engineering and services support.

CHAPTER FIVE

Forward Support Location/CONUS Support Location Preparation for Meeting Uncertain FOL Requirements

The ability to quickly link a global network of forward support locations (FSLs) and CONUS support locations (CSLs) to meet FOL deployment and sustainment needs is vital to every operation. In this chapter, we analyze data from JTF NA, OEF, and OIF to illustrate the importance of this global network. We also discuss some of the limitations of the current network in meeting operational needs.

Findings

Our findings are in the following areas:

- FSLs as supply locations
- CSLs and C2
- Maintenance FSLs/CIRFs.

FSLs as Supply Locations

In all three recent operations, combat support resources dominated the footprint at the FOLs, as shown in Figure 5.1. An analysis of the JTF NA, OEF, and OIF TPFDDs and/or execution data, as well as of data provided by DynCorp, the CENTAF WRM contractor for OEF and OIF, shows that aviation units and their associated maintenance functions accounted for only 20 percent of the tonnage moved

Figure 5.1
FOL Footprint

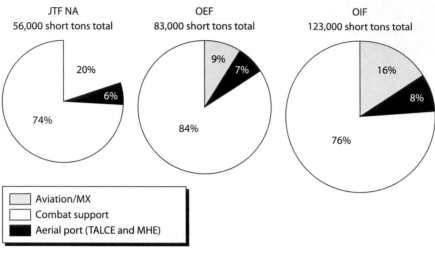

to FOLs in JTF NA, only 9 percent during OEF, and only 16 percent during OIF. Aerial port equipment[1] accounted for 6 percent during JTF NA, 7 percent during OEF, and 8 percent during OIF. The remaining 74 percent for JTF NA, 84 percent for OEF, and 76 percent for OIF consisted of combat support resources.[2] In total, approximately 123,000 short tons of materiel and personnel were moved in support of OIF.

An analysis of the combat support portion of the OEF TPFDD, the OIF execution data, and DynCorp data showed that 68 percent of the OEF requirement and 35 percent of the OIF requirement were FOL support (see Figure 5.2). The term *FOL support* is used to identify the base operating support—that is, those resources that are

---

[1] *Aerial port equipment* includes Tactical Airlift Control Element (TALCE) and associated items and equipment, such as material handling equipment (MHE), that enable strategic and tactical airlift operations.

[2] *Combat support* consists of civil engineering, communications, security forces, maintenance, service, munitions—anything other than the actual flying operation.

Figure 5.2
Combat Support Footprint

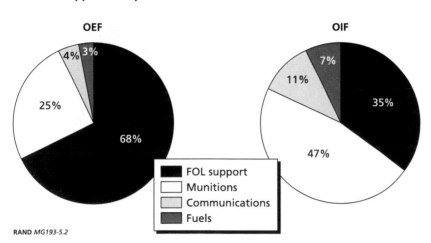

RAND *MG193-5.2*

required to set up and sustain a base. This resource category includes, but is not limited to, civil engineering equipment; WRM, including tentage, shower/shave, and water-purification systems; and vehicles. Also included in this category are the industrial and kitchen sets that round out the base support packages.

During JTF NA, WRM was distributed from Sanem, Luxembourg, and CONUS. During OEF and OIF, WRM was distributed from locations in the Middle East, Europe, and CONUS.

Combat support resources include munitions, communications, and fuels support. Munitions resources made up 25 percent of the combat support requirement during OEF and 47 percent of the total combat support requirements during OIF. Communications equipment accounted for 4 percent during OEF and 11 percent during OIF. FMSE—for example, bladders, hoses, pumps, not fuel itself—made up 3 percent of the movement during OEF and 7 percent of the movement during OIF. Note that these percentages include only those items listed in the TPFDD/execution plan and documented by DynCorp; they do not include food, water, and fuel.

During JTF NA, FSLs and CSLs satisfied approximately 76 percent of the combat support requirements. During OEF, FSLs satisfied the largest portion of the combat support requirement (see Figure 5.3), approximately 64 percent. FSLs provided approximately 82 percent of the total FOL support resources needed during OEF. Another 16 percent was munitions-related. During OIF, FSLs met 77 percent of the combat support requirement. Of that materiel, 40 percent was FOL support and 56 percent was munitions.

During OEF, CSLs also satisfied a portion of the FOL requirements, although a much smaller portion of the overall support—only approximately 11 percent. Of the combat support resources moved from CONUS, only 13 percent was FOL support. Most of the CONUS support—approximately 85 percent—was munitions-related. During OIF, CSLs satisfied only 6 percent of the combat support requirements. Of the materiel moved from CONUS, only 16 percent of the requirement was FOL support; 61 percent was munitions-related.

Although providing the majority of the FOL total resource requirements during OEF and OIF, FSLs did operate with some constraints. During OEF, several resource constraints were noted; many of them still existed during OIF. Many of the FSLs were located at or

Figure 5.3
**Combat Support Requirements Were Resourced Mainly
from FSLs During OEF and OIF**

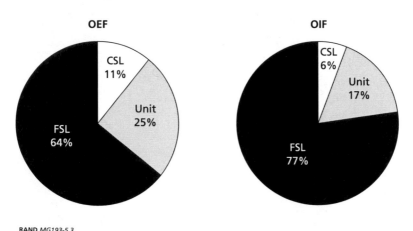

near places the Air Force intended to use as FOLs—for example, Al Udeid and Thumrait Air Bases. Moving bombers or tankers into a location while moving support equipment out, created ramp space- and equipment-utilization issues. Fortunately, the long timelines involved in readying the AOR for OIF mitigated some of these issues. However, workload and beddown requirements at these locations could create conflicts. During OEF, some sites completely stopped the outload of equipment while the contractor teams helped to build tent cities before force packages arrived at the combined FSL/FOL location. This vying for contractor resources could adversely affect deployment timelines to other FOLs where outloads from the FSLs are needed to provide the equipment that allows bases to be set up and operations to begin. Additionally, ramp space that is consumed by aircraft operating from the combined FSL/FOL site is not available for airlift aircraft to move equipment out of the FSL. However, even with these problems, FSLs satisfied most FOL resource needs.

CSLs and C2

During JTF NA, CSL resource constraints hindered CSL effectiveness. Specifically, backorders added substantial resupply time and variability during the conflict. Prioritization of supply resulted in an unequal readiness level in CONUS and across the rest of the Air Force—an undesirable side effect of prioritizing scarce resources. Although backorder rates improved, they remained high throughout JTF NA.

CSLs were used more effectively during OEF. Because of the experiences in JTF NA, attention was given to creating better links between CSLs and the warfighters. To enhance CSL responsiveness to the warfighter, as discussed in Chapter Three, AFMC tasked the Logistics Support Office, Headquarters, AFMC, to monitor shipment pipelines and track the delivery times to various locations by various commercial and military transportation modes. Delivery-time information was relayed to customers so that they could make better decisions about transportation modes for future shipments.

AFMC/LSO also developed a website with estimated shipping times and best methods for shipping to different locations. The website was updated to show any anomalies in shipping caused by customs problems or host-nation restrictions so that alternate routing could be used.

Air Force Materiel Command, along with its customer MAJCOMs, also created the High Impact Target list on which MAJCOMs identified their most important repair parts for AFMC to monitor in the various Air Logistics Centers. This program is popular with the customer MAJCOMs. AFMC has now automated many of the processes associated with maintaining the list and gathering status reports. See Chapter Three of this report for more details on the TO-BE CSC2 operational architecture.

As a result of experiences during OEF, the AFMC/LG created the Warfighter Sustainment Division (WSD) at Headquarters, AFMC, as the single focal point for the warfighter at AFMC. The WSD now manages the HIT list, as well as providing rapid, 24-hour, 7-days-a-week logistics support for all concerns for AFMC (Air Force Materiel Command, 2003).

Maintenance FSLs

Another success was the use of centralized intermediate repair facilities to satisfy a range of intermediate repair operations for fighter units deployed to Operation Northern Watch, Operation Southern Watch, in support of OEF, and during OIF.

During JTF NA, the use of CIRFs, also called maintenance FSLs, was successful in meeting the warfighters' needs. Three existing U.S. Air Force, Europe, FSLs were formally designated as CIRFs during JTF NA: RAF Lakenheath, England; Aviano AB, Italy; and Spangdahlem AB, Germany. RAF Mildenhall, England, was later developed as a CIRF to support tankers. JTF NA showed that preselection and resourcing of CIRFs can improve flexibility and reduce deployment footprint.

As a result of successfully using CIRFs on an ad hoc basis during JTF NA, the Air Force developed and tested a CIRF Concept of Operations (CONOPS) for supporting the AEF. The Air Force CIRF test began in September 2001. The CIRF CONOPS was adjusted to include support to OEF forces once those operations began.

Established in USAFE to support forces deployed to Operation Northern Watch and Operation Southern Watch prior to OEF, CIRFs satisfied a range of intermediate repair operations for fighter units deployed to Operation Northern Watch, Operation Southern Watch, and OEF. The repair facilities at RAF Lakenheath were identified for F-15 line replaceable unit (LRU) repair, as well as for LANTIRN pods and F-100 engines. Spangdahlem Air Base in Germany was designated as the repair facility for ALQ-131 electronic countermeasure (ECM) pods and F-110 engines. Later, when a backlog developed at Lakenheath, USAFE added the LANTIRN repair facility at Aviano Air Base in Italy.

The transportation segment that had been planned to support Operation Northern Watch and Operation Southern Watch was expanded to include the other OEF locations. Plans were developed to move items from forward bases to Operation Northern Watch/ Operation Southern Watch locations for onward movement (the theater distribution system will be covered in more detail in Chapter Six).

As a result of successfully using CIRFs on an ad hoc basis during JTF NA and of successfully testing the CIRF CONOPS during OEF, CIRFs established in USAFE to support forces deployed to Operation Northern Watch and Operation Southern Watch prior to OEF were used again in OIF. Before OIF was initiated, the CIRF CSC2 element was stood down. Maintenance personnel were moved back to their functional duties in the Headquarters, USAFE, staff. As OIF began, the CIRF CONOPS was adjusted again to include support to OIF forces (see Figure 5.4). Existing CIRFs had funding procedures in place and in operation. Ramstein AB, Germany, became the CIRF for wheels, tires, brakes, and C-130 engines during OIF. Since operations were quite short in Iraq, CIRFs were never stressed during combat operations. The ease with which the CIRFs operated should be attributed to having a well-thought-out concept of operations and an executable plan. However, if combat had been longer, transportation might have become an issue.

The CIRFs reduced the deployment requirement during OIF in both personnel and in materiel, as shown in Figure 5.5. For example,

Figure 5.4
CIRFs Provided Maintenance Support for Fighters During OIF

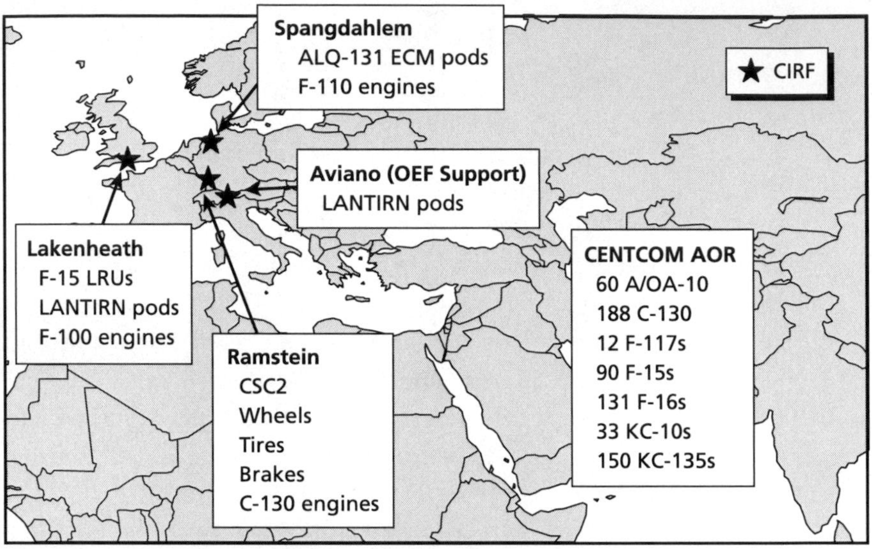

Spangdahlem
ALQ-131 ECM pods
F-110 engines

★ CIRF

Aviano (OEF Support)
LANTIRN pods

Lakenheath
F-15 LRUs
LANTIRN pods
F-100 engines

CENTCOM AOR
60 A/OA-10
188 C-130
12 F-117s
90 F-15s
131 F-16s
33 KC-10s
150 KC-135s

Ramstein
CSC2
Wheels
Tires
Brakes
C-130 engines

RAND *MG193-5.4*

the CIRF was able to support all Southwest Asia (SWA) repair needs for ALQ-131 pods with the existing equipment at Spangdahlem and nine additional personnel. If this repair capability were deployed forward instead, then each deployed unit would need seven personnel and 13 short tons of support equipment. With 13 deployed units, these requirements would total 91 personnel and 169 short tons of equipment. Similar savings were achieved for Jet Engine Intermediate Maintenance (JEIM), LANTIRN, and avionics intermediate-maintenance shop (AIS) personnel and equipment. These three CIRFs combined to save the deployment of 133 personnel and almost 247 short tons of support equipment to FOLs, which would have been required under a decentralized structure.[3]

The reduced footprint resulting from CIRFs freed a significant amount of deployment airlift. However, for CIRFs to operate in an expeditionary manner, assured sustainment transportation, beginning

---

[3] Execution data were abstracted from JOPES, April 24, 2003.

Figure 5.5
CIRFs Reduced the Southwest Asia/AOR Footprint

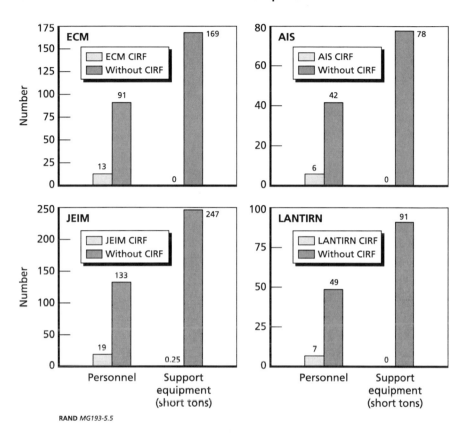

on Day 1, needs to be provided. CIRFs need to articulate their TDS requirements during planning processes (Peltz et al., 2000).

Implications

A review of recent operations indicates that future conflicts are likely to occur far from CONUS. A global network of FSLs with pre-positioned WRM is necessary to meet AEF goals. The use of austere FOLs and an immature theater infrastructure in both OEF and OIF has illustrated the need for a portfolio of FSLs. The current AEF force structure of light, lean, and lethal response forces is highly dependent on FSLs.

When developing a portfolio of FSLs to support numerous different operational challenges, the Air Force needs to ensure that it provides many options and that those options are available for use in future contingencies. Trade-offs need to be examined between existing FSLs whose improvement may enhance throughput and storage capacity and capabilities that can be developed by investing in new FSLs in differing locations.

When considering whether to develop new FSLs or improve existing facilities, attention should be given to joint requirements. All military services depend upon materiel that has been prepositioned to meet contingency requirements. The management of joint facilities to meet multiple service requirements may reduce operating costs. Information should be shared among services as well as with U.S. allies. If such arrangements are pursued, throughput required for all participants should be explicitly considered.

Since the centralized intermediate repair facility CONOPS has been successful in the past two operations, CIRFs will likely be more widely used in future operations. As CIRFs are used in more operations, their requirements for reliable transportation should be included in the planning process. The trade-off of reduced deployment airlift in the early stages of a conflict is the availability of reliable sustainment transportation beginning on Day 1 of the operation. Without assured airlift, CIRFs will struggle to meet AEF operational goals. More work is required to ensure that the combatant commanders understand and support the risk the Air Force is taking when agreeing to maintain aircraft using CIRF.

CHAPTER SIX

Reliable Transportation to Meet FOL Needs

Without a reliable transportation system, deployment can be delayed and sustainment can be hindered. In this chapter, we discuss transportation and movement experiences in JTF NA, OEF, and OIF.

Findings

Our findings are in the following areas:

- Movement by commodity
- Modes of transportation
    —Munitions
    —FOL support assets
    —Spares
- Management of the theater distribution system (TDS)
    —TDS responsibility and organization
    —In-transit visibility.

Movement by Commodity

During all three operations, JTF NA, OEF, and OIF, the TDS movement was dominated by such commodities as fuel,[1] munitions,[2]

---

[1] Fuels data are from CSC Fuels Consumption Log, July 2002 and April 2003.

[2] Munitions data are from CENTAF Munitions Expenditure Rollup, 15 Mar 02, and execution data abstracted from JOPES, April 24, 2003.

FOL support, rations,[3] and spares.[4] These are bulky commodities, and their movement requires large transportation capacity. Figure 6.1 shows the tons of items that were moved during the first 100 days of OEF and during OIF.

As shown in Figure 6.1, fuel dominated combat support commodities that were moved. Fuel such as JP8 jet fuel was not always available at contingency locations; therefore, it was moved by all modes of transportation—airlift, sealift, on trucks over land, and by direct delivery from pipelines. Nonstandard, host-nation fuel, which had been used more frequently in recent contingencies, is included in the total fuel movement.[5] All modes of transportation of fuel are cal-

Figure 6.1
Commodities That Were Moved During OEF and OIF

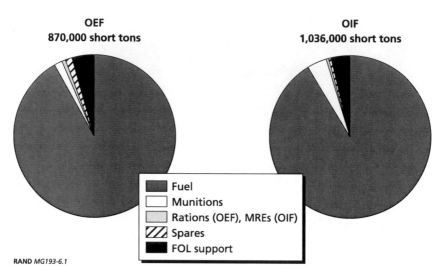

---

[3] Rations data are from CENTAF/A-4 LGV and U.S. Air Force, Central Command (2003).

[4] Spares data are from AFMC/LSO-LOT.

[5] Given the availability of nonstandard fuels in remote, Third World locations, the Air Force needs to understand the different properties of the host-nation nonstandard fuel and additives being used during contingencies.

culated in the total fuel-movement requirement shown in the figure. The total fuel consumed during OEF was approximately 800,000 short tons; during OIF, it was approximately 945,000 short tons.[6]

Having an Air Force Petroleum Office–deployed area lab in the AOR provides immediate analysis of fuel and oxygen samples increasing mission responsiveness while decreasing transportation time. Having the lab in the AOR reduced the in-transit time of samples. For example, transportation time from Iraq to Al Udeid was only three days as opposed to three weeks from Iraq to Mildenhall, United Kingdom.

Commodity movements in support of OEF and OIF, other than fuel, are broken out in Figure 6.2. The total amount of rations moved during OEF was approximately 4,000 short tons. Approximately 3,000 short tons of meals, ready-to-eat (MREs) were moved

Figure 6.2
Other Commodity Movement During OEF and OIF

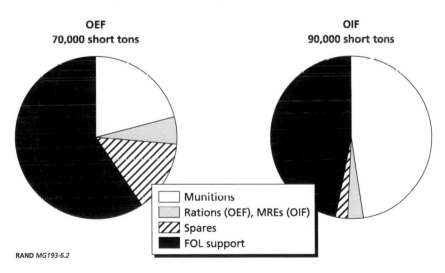

RAND MG193-6.2

---

[6] Fuel consumption is calculated on the basis of fuel consumed during March and April 2003.

during OIF. Other important assets include munitions and FOL support. Taken together, both accounted for over 90 percent of the movement. During OEF, FOL Support accounted for 42,000 short tons of movement. During OIF, the FOL support moved was only slightly larger, at 43,000 short tons. Munitions moved in support of OEF were 15,000 short tons. During OIF, munitions accounted for almost 44,000 short tons.

Finally, spare parts were included in the sustainment movements. Spares are a small portion of the movement (approximately 10,000 short tons during OEF and only approximately 2,000 short tons during OIF), but they are critical to weapons system support.

OEF Modes of Transportation

The Air Force uses many modes of transportation and different commodity supply chains to move large amounts of materiel in order to initiate and sustain combat operations. Figure 6.3 illustrates the modes of transportation used during OEF to move combat support materiel only, excluding aviation, maintenance, or aerial port materiel.[7] Intratheater airlift consisted of approximately 15,400 short tons.[8] Approximately 11,000 short tons were moved by sea,[9] 34,000 short tons were moved by land,[10] and 1,200 short tons were moved by intertheater airlift.[11]

OEF Munitions. Combat aircraft expended over 7,000 tons of munitions during OEF, a good portion of which was precision-guided. As with fuels, munitions were moved using various modes of

---

[7] This 67,000 short tons does not include the 3,000 short tons of rations moved in support of OEF.

[8] Data were abstracted from OEF TPFDD.

[9] Data were provided by Ammunition Control Point (ACP).

[10] Data were provided by DynCorp.

[11] Data were abstracted from OEF TPFDD.

Figure 6.3
Several Modes of Transportation Were Used to Move Commodities During OEF

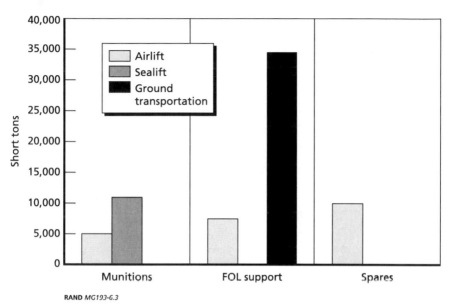

RAND *MG193-6.3*

transportation. The total munitions moved were approximately 16,000 tons, including approximately 5,000 tons airlifted in the form of Standard Air Munitions Package (STAMP) packages, 3,000 tons moved by sea from CONUS, 8,000 tons used from aboard APF ships, and a few hundred tons as part of the bomber deployments themselves.

During OEF, the British protectorate of Diego Garcia served the primary Air Force bomber FOL. Diego Garcia is one of four FOLs around the world identified for use by Air Force heavy bombers in the event of crisis. These so-called bomber islands are places where significant infrastructure and reserves of materiel can be built up ahead of time. JTF NA's experience highlighted the importance for such bomber islands. However, in 2001, Diego Garcia was still under development. The Air Force was in the process of stocking munitions at Diego Garcia. Additional munitions were brought into Diego Garcia from CONUS during the buildup phase, just before the start of bombing operations. These munitions included JDAM kits and Wind Corrected Munitions Dispensers (WCMDs), as well as the heavy bomb bodies themselves. But, OEF operations began before the stocking was complete.[12]

The Air Force maintains munitions inventories aboard the three ships of the Air Force Afloat Prepositioning Fleet. Usually, these ships are deployed forward in different regions of the world. However, at the time of Operation Enduring Freedom, one of the ships, the MV *Buffalo Soldier*, was being off-loaded in CONUS. Its cargo, normally stored in bulk format, was being transferred to containers for storage aboard the MV *A1C William H. Pitsenbarger*. As a result, the Air Force had only two of its APF ships deployed forward, which led to some reluctance on the part of the Air Force to release APF assets for OEF. In addition to using APF assets from the MV *MAJ Bernard F. Fisher*, the Air Force contracted the sealift vessel *Cornhusker State* to bring assets from the *Buffalo Soldier*, as well as other assets destined for the *Pitsenbarger*, to Diego Garcia. Sealift delivered large quantities of munitions. However, it took approximately 28 days for the *Cornhusker State* to sail from the East Coast to Diego Garcia.[13] Although it took longer to have the munitions moved by sea than it would have to move them by air, with enough time, munitions were in place when needed.

OEF FOL Support Assets. While munitions were moved primarily by air and sea, FOL support assets were moved mainly by ground transportation during OEF (see Figure 6.3). Most of the FOL support came from forward support locations in the area of responsibility; however, some of it was transported by air from CONUS.

Delivery of FOL support assets was sometimes faster for equipment coming from CONUS than for equipment coming from within the AOR. While deliveries to FOLs from FSLs in the same country were quick, an average of approximately 4 days, FOL support trans-

---

[12] Data are from HQ USAF/IL External Slides from the daily briefings dated September 19 and 20, 2001.

[13] Data were provided by ACP.

portation times to FOLs from FSLs in another country could be much slower, ranging from two to five weeks. In contrast, FOL support deliveries originating from CONUS closed in four to 15 days.[14] Air transportation from CONUS and FSLs in the same country was the quickest method of FOL support delivery.

Many reasons can be given for the slow FSL intercountry delivery closing times—for example, limited TDS capacity, slow WRM warehouse throughput, or problems getting agreement of the receiving host nation. In one case, the receiving base did not have the personnel to construct FOL assets, so the base leadership requested that assets be held at the FSL. Intratheater airlift, especially in the early days of OEF, was in extremely short supply, with only a few C-130s in-theater. The lack of cargo aircraft was not because of a lack of airlifters in the fleet but because of a lack of beddown space at the various FSLs, which were also serving as combat, ISR, and tanker bases.

Line-haul trucks were contracted locally but were subject to availability, road conditions, and, for some sites, the availability of ferries. In some cases, Air Force spare parts piggy-backed on trucks that were contracted by the WRM contractor (DynCorp) to carry FOL support equipment. Locally contracted trucks presented a force-protection concern, requiring additional inspections, escorts, or transloading, all of which require additional time.

OEF Spares. Spare parts are a small but vital part of the sustainment movement. Spares accounted for only 1 percent of the total sustainment movement during OEF;[15] however, OEF success depended on the availability of the Air Force's high-demand/low-density assets. The small fleet sizes of ISR assets (U-2, Predator, Global Hawk, and E-3) and AFSOC fixed- and rotary-wing aircraft demanded immediate spare parts support from CONUS. Movement of spares, especially movement by air, depends upon the theater distribution network.

---

[14] Data were provided by CENTAF.

[15] Data were provided by AFMC/LSO-LOT.

No single source of air transportation was best for every destination where spare parts were involved. Transportation-time data during OEF shows that, at some locations, commercial carriers were faster, whereas, at other locations, AMC was faster. Furthermore, the performance of AMC relative to that of commercial carriers, and even that among the different commercial carriers, varied from week to week.

Figure 6.4 displays some spares transportation data collected by the Air Force and the RAND SDMI project from October through December 2001. The small squares on the graph show the mean military airlift (MILAIR) times and mean World Wide Express (WWX) airlift times of items to several locations in Southwest Asia in support of OEF. Also shown on the figure are the median, the 75th percentile, and the 95th percentile.

OIF Modes of Transportation

Just as in OEF, the Air Force used many modes of transportation and different commodity supply chains to support Operation Iraqi Freedom. Figure 6.5 illustrates the modes of transportation used during OIF to move combat support materiel. These totals do not include aviation, maintenance, or aerial port materiel.[16] Airlift consisted of approximately 19,000 short tons.[17] Approximately 39,000 short tons were moved by sea,[18] and 28,000 short tons were moved by land.[19] Sealift was used much more heavily in OIF than in OEF.

   OIF Munitions. Approximately 44,000 short tons of munitions were moved in support of OIF. Multiple means transported these munitions to the AOR, including airlift of approximately 3,200 tons, sealift of 500 short tons from CONUS, sealift of 1,400 short tons

---

[16] These 87,000 short tons do not include the 3,000 short tons of MREs moved in support of OIF.

[17] Execution data were abstracted from JOPES, April 24, 2003.

[18] Data provided by HQ AF/IL ACES.

[19] Data were provided by DynCorp.

Figure 6.4
No One System Is Best in All Cases, as the Performance of MILAIR and WWX Indicate for Air Transportation of Spares During OEF

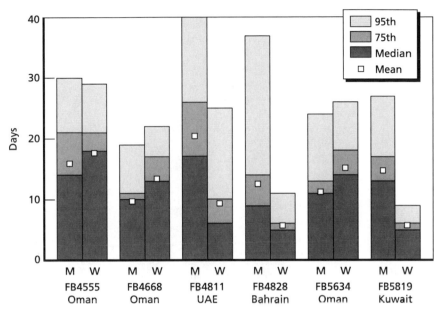

SOURCE: Data are from the SDMI database, run February 6, 2003. An item code (FBxxxx) is listed above the location name for each item in the figure.

NOTES: In these data, airlift times begin when the cargo leaves the CONUS Defense Logistics Agency (DLA) distribution center and ends when the cargo receipt paperwork has been completed by the Transportation Management Office or base supply. M=military airlift; W=World Wide Express (WWX) airlift.

RAND MG193-6.4

from FSLs, and 29,000 short tons used from aboard APF ships. Also, 10,000 short tons were already in place in the AOR, and a few hundred tons were part of the bomber deployments themselves.

Most of these munitions were moved on APF. Because of the long planning time for OIF, the Air Force was able to take advantage of several sealift vessels. Prior to OIF, one ship was docked in CONUS, for hull recertification. While that work was being completed, the Air Force had extra munitions loaded on the ship. The Air Force saved approximately 100 C-5s' worth of airlift by loading the extra munitions on the ship.

Figure 6.5
Several Modes of Transportation Were Used to Move Commodities During OIF

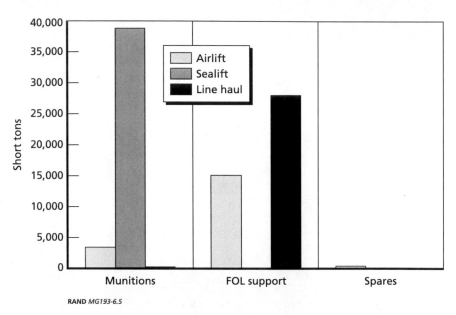

RAND *MG193-6.5*

The Air Force also took advantage of an Army prepositioning ship to move munitions from CONUS to the AOR. The Air Force added containers to the top deck of the Army ship, again saving valuable airlift. Ships were also used to transport munitions within the AOR. For example, munitions were moved from Saudi Arabia to FOLs in the AOR by ship.

Normal sealift from CONUS to the AOR, including loading and off-loading, takes about 71 days compared with 8 to 12 days by airlift. Because time was not a factor, the Air Force could send munitions by sea to the AOR and not have to compete for valuable and constrained airlift at the time of execution.

OIF FOL Support Assets. While munitions were moved primarily by sea, forward operating location support assets were moved primarily by air and by ground transportation. Most of the FOL support came from forward support locations in the AOR; however, some was transported by air from CONUS.

Support deliveries to FOLs from FSLs in the same country were quick. Support deliveries to FOLs from FSLs in another country could be much slower. The same issues that were encountered during OEF were again encountered in OIF. Limited TDS capacity and WRM warehouse throughput, and difficulty obtaining host-nation agreements caused delays in intercountry transportation. Lack of cargo aircraft and a lack of ramp space also caused delays.

As in OEF, line-haul trucks were contracted locally but were subject to availability and road conditions. Many of the FSLs used during OEF were again used during OIF. Air Force personnel had experience working transportation issues, so cargo was moved quickly when necessary.

OIF Spares. As stated earlier in this chapter, spare parts are a small but vital part of sustainment movement. Spares accounted for only approximately 2,000 short tons of sustainment movement during OIF.[20] As in OEF, the Air Force's high-demand/low-density assets (for example, U-2, Predator, Global Hawk, and E-3) demanded immediate spare parts support from CONUS. The mode of transportation depended on the requirement. Spares had to arrive quickly; therefore, they were airlifted.

For example, the Deployable Air Traffic Control and Landing Systems (DATCALS) enabled all-weather flying in the AOR during OIF. As key Iraqi airfields were liberated, new DATCALS requirements were established without any significant reduction in OEF DATCALS commitments—resulting in the largest deployment of DATCALS in Air Force history. Deploying the majority of the fleet posed many logistical challenges; delivering scarce spare parts was the first priority.

Air Combat Command partnered with the DATCALS depot and leveraged $3.2 million of Global War on Terrorism funds to dramatically improve the priority for DATCALS repair. While this effort immediately reduced delay times for the most-pressing mission-capability requisitions, it also eventually replenished the on-site spares

---

[20] Data were provided by AFMC/LSO-LOT.

kits at each deployed location and minimized future outages. The leveraging of spare parts proved invaluable in the successful employment and sustainment of DATCALS during OEF and OIF.

As in OEF, no single source of transportation was best for every destination where the movement of spare parts was concerned. Figure6.6 shows data collected by SDMI for April 2003. The small squares on the graph show the average military airlift (M) times and average WWX (W) times[21] to several locations in Southwest Asia in support of OIF.[22] The data show that one mode of airlift was not

Figure 6.6
No One System Is Best in All Cases, as MILAIR and WWX Indicate for Air Transportation of Spares During OIF

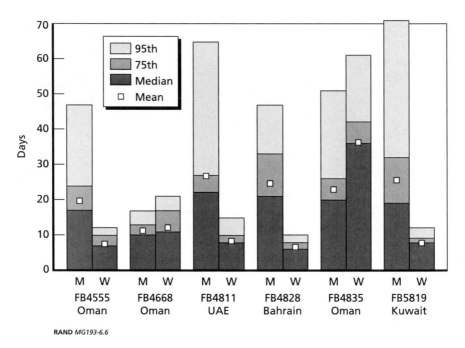

RAND *MG193-6.6*

---

[21] These data are based on the transportation time starting when the cargo is requisitioned and ending when the cargo receipt paperwork has been completed by the Transportation Management Office or base supply.

[22] Data from April 2003 DODAAC report from SDMI.

consistently faster. For some locations, commercial carriers were faster; for others, AMC was faster. Furthermore, the relative performance varied from week to week.

Management of Theater Distribution

Since no one mode of transportation dominates all locations, those charged with ensuring prompt resupply of materiel need to select the mode that best meets their needs. As experienced in JTF NA, multiple supply chains need to be used to ensure responsive delivery to warfighters at different locations. To make proper use of these different supply channels, planners must have ready access to information necessary for them to make good transportation decisions.

The theater distribution system has two main responsibilities. The first deals with moving assets from the FSLs to the FOLs. This part of TDS is required to move initial deployment and sustainment items to where the items are needed, many of which are stored at or near the AOR. The second part of TDS, a tactical distribution system, provides the onward movement of resources from CONUS and the movement of reparable parts to and from FSLs.[23]

There were coordination problems and gaps between the TDS and the strategic movements system during JTF NA and then again during OEF. TDS was slow to evolve, and intertheater and intratheater movements were not well coordinated. Many problems arose in establishing a theater distribution system to meet Air Force needs. They began with the Air Force playing a larger role in the development and design of the TDS than expected.

With the experience provided by OEF, the theater distribution system was better prepared for OIF. Since many of the FOLs and FSLs used during OEF were again used during OIF, routes were already established before OIF even began. However, some gaps between the two systems, strategic movement and theater movement, still existed. Cargo piled up at transshipment points, and there were

---

[23] The Secretary of Defense named USTRANSCOM the Department of Defense Distribution Process Owner on September 16, 2003. USTRANSCOM will be responsible for synchronizing global and theater distribution processes.

problems identifying priority movements between services. Other issues included en route refueling. Doctrine does not indicate who is to provide the assets and personnel necessary to support a bare base en route fueling mission. Although problems still existed in coordinating the TDS, execution appeared to be better than during JTF NA or OEF.

Another transportation issue, involving payment for shipping, arose after major combat operations were completed in OIF. A difference in how transportation shipments are billed led to the use of airlift when it may not have been necessary. Strategic air is paid for when used, from an industrial fund. Theater air shipments are "free" to the theater shippers; theater shippers do not have to pay for theater airlift. Commercial trucks contracted must be paid on an as-used basis by the shipper. This difference in pricing services can cause a misallocation of air assets. For example, some cargo that may be better moved by surface may be placed for air delivery, since air delivery is free for the shipper.[24] Although this problem arose after major combat operations were over, it is another systemic problem with the theater distribution system.

TDS Responsibility and Organization. According to doctrine, the combatant commander designates which service will have responsibility for the Joint Movement Center[25] and TDS. This service would be responsible for the planning and execution of all movements of materiel and personnel within the AOR by land (trucks and rail), sea (ships and barges), and air. The combatant commander may designate the service that is most capable of performing the tasks or the predominant user of the system to be responsible for developing the theater distribution system (Joint Chiefs of Staff, 2002).

In past operations, development of the TDS was an Army responsibility. At the beginning of JTF NA, the air component had the preponderance of forces; therefore, the Air Force was given responsi-

---

[24] Telephone interview with Maj Gen Robert Elder, Central Command, Deputy CFACC, August 20, 2003.

[25] The Joint Movement Center is responsible for validating and prioritizing air mobility requirements.

bility for TDS. The same situation occurred during OEF. At the start of OEF, the Army had few forces in theater and resupply to geographically dispersed, austere areas was largely by air. Consequently, responsibility for the JMC and TDS was delegated to the Air Force through the AFFOR A-4 (who also acted as the CFACC C-4) by Central Command. As OIF planning unfolded, a theater distribution system was designed. A key difference was that TDS was developed before operations began, using the TDS already in place for OEF as a guide.

TDS is vitally important for meeting rapid deployment and resupply needs of the AEF. In JTF NA, as in OEF, many problems arose in establishing a TDS to meet Air Force needs. In JTF NA, these problems began with the Air Force playing a larger role in the development and design of the TDS than had been anticipated. Air Force personnel assigned to this responsibility may have neither the training nor the background necessary to develop a TDS.

The theater distribution system used during OEF evolved over time. The Joint Movement Center was collocated with the Air Mobility Division (AMD) in the CAOC, a situation that did help in planning and coordinating air movements. However, the JMC and TDS involve *all* modes of transport.

CAOC personnel working TDS had a difficult time projecting distribution system requirements. This same problem existed in JTF NA. Unable to come up with good estimates for the TDS, the initial OEF TDS relied on the TDS in-place—four C-130s—to support Operation Southern Watch. This capacity proved inadequate to meet OEF TDS needs; a few months into OEF, large backlogs of cargo developed at transshipment points in the AOR (HQ USAF, 2001b; Barthold, 2002, p. 5). The backlog peaked at 1,000 pallets and persisted during the first 100 days of the operations.[26] It was not until several months into the operation that standard air routes (STARs) were established. TDS was further complicated by Operation Southern Watch, which was ongoing in the same theater at the same time.

---

[26] Two hundred pallets were considered an acceptable backlog. A slide with backlog data is available for review upon request.

Prioritization among different ATOs and associated FOLs was difficult.

During OIF, there were coordination problems and gaps between the TDS and the strategic movements system. The intratheater distribution system, however, was better organized in OIF than the system in OEF. STARs were established before combat operations began. A long chain of command allocated adequate airlift for meeting airlift to the AOR. TDS was not developed in the ad hoc manner in which it had been developed during OEF. TDS was established early and, on the surface, appeared to function well. However, intertheater and intratheater movements were still not well coordinated.

Air Force personnel working TDS had a difficult time projecting distribution system requirements, even after initial assessments were made. This same problem existed during JTF NA and OEF. While attention was given to monitoring how well the system was performing, early decisions were made about the extent of backlog that would constitute "adequate performance." The existing TDS capacity was inadequate to meet OIF transportation needs. Backlogs of parts awaiting shipment to CIRF built up, and movement of critical communications equipment was delayed.

Figure 6.7, which provides a sampling of four bases, gives a breakdown, by segment, for Air Force shipments in support of OIF during April 2003.[27] The first segment, "Ship to POE," represents the time it took, once an order was received, to ship materiel to the port of embarkation (POE). Materiel spent a significant portion of time waiting to be shipped. "POE Hold" shows how long materiel sat at the POE. The next segment, "POE to POD," shows the transportation time required to move materiel from the POE to the port of debarkation (POD). "POD Hold" represents the time materiel sat at the POD awaiting transportation to the FOL or FSL. The time spent at the hubs (POE and POD) includes the time that cargo waited for

---

[27] Data are from the April 2003 DODAAC report from SDMI.

Figure 6.7
Backlogs at Transshipment Hubs During OIF Varied Widely

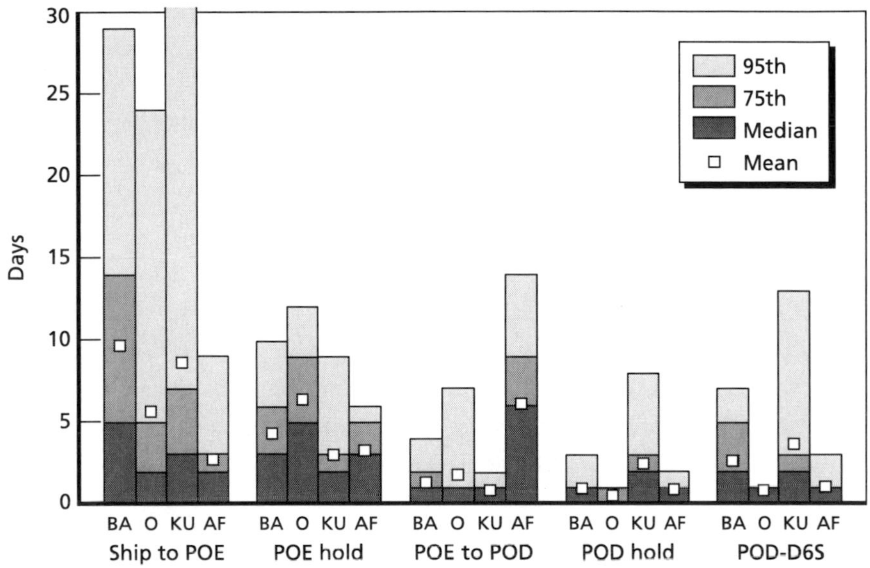

NOTES: D6S = receipt at the base; BA = Bahrain; O = Oman; KU = Kuwait;
AF = Afghanistan.

RAND MG193-6.7

the TDS to deliver cargo from transshipment points after being downloaded by the strategic movements system. (These hold times could be a function of combatant commander priorities and not necessarily how fast the theater distribution system could move if the materiel had a high priority.) And finally, POD to D6S represents the transportation time from the POD to receipt at the base. Another TDS issue was each service's different prioritization rules. Whereas the Air Force places the greatest value on high-tech parts, the Army places first priority on personal items—an inherent conflict.

In-Transit Visibility. In-transit visibility (ITV) was a problem during OEF. After Operation Desert Storm, transporters had to track ITV on only four C-130s in the Central Command AOR.[28] With the

---

[28] Interview with Maj Gen Richard Mentemeyer, October 2002.

beginning of OEF and more aircraft in the AOR, ITV became much more difficult to track. Once units and individual personnel[29] left their home stations, ITV on them was often lost.[30] Visibility was lost at transshipment points, such as Rhein Main, where large shipments were subdivided into smaller shipments going many places.

Ramstein had ITV difficulties because of the sheer volume of materiel and personnel passing through the base. New software, Deployable Global Air Transportation Execution System (DGATES), was put in place at the beginning of OEF. The Air Force accelerated the installation of DGATES to assist with ITV. Reportedly, data from Global Air Transportation Execution System (GATES) were useful, but DGATES had some problems.[31] Because DGATES and other systems rely on human input, the personnel inputting the data need to understand the importance of the data and the result of inaccurate and incomplete data. For example, during OEF, units would arrive at the aircraft without proper documentation. Even without the proper paperwork,[32] the units would be shipped, which could cause a loss of visibility for the shipment.

ITV proved to be problematic during OIF as well. For example, the Aerial Bulk Fuel Delivery System was designed to use at least two 3K fuel bladders. The C-130 modifications (armoring) allow only one 3K bladder to fly per aircraft, splitting up the delivery system package. In addition, the delivery system was treated as cargo and off-loaded along with fuel, making it difficult to track.

There continued to be problems in establishing in-transit visibility when shipments moved from one system to the other—that is, from strategic to TDS. Each MAJCOM had teams to monitor JOPES and other systems to ensure ITV. In addition, the Warfighter

---

[29] ITV was available only on Civil Reserve Air Fleet (CRAF) and MILAIR transport. Commercial airlines are not required to share manifest data with the military.

[30] Col Bruce R. Barthold, "Major Issues from the AFCESA/AFIT Sponsored Operation Enduring Freedom RED HORSE and Prime BEEF Lessons Learned Conference, 13–15 Nov 02," memorandum dated December 2002, p. 4.

[31] Interview with Mr. Frank Weber, USTRANSCOM J3/J4, May 2002.

[32] Interview with Mr. Frank Weber, USTRANSCOM J3/J4, May 2002.

Sustainment Division at AFMC monitored ITV for shipments into and out of the depots or other sources of supply. AMC has designated AMC/A4TI as the lead agency to be their MAJCOM advocate for ITV-related issues, to include an ITV cell manned 24 hours a day, seven days a week. USTRANSCOM is the Department of Defense (DoD) proponent for ITV and, as such, has the ability to initiate an ITV division or cell to identify and better track worldwide ITV.

Implications

If the Air Force is asked to be responsible for TDS, as it was in JTF NA and OEF, or even if it just provides input to another service that controls TDS, as was the case in OIF, the Air Force needs to provide education and training to handle the TDS responsibilities. Creation of a logistics readiness officer shows promise for fulfilling this critical need. However, the Air Force should develop a specific education and training plan for theater distribution.

The transportation system used during any operation will be complex, multimodal, and involve numerous customers (for example, Army, coalition, and Air Force). Theater distribution is more than just the onward movement of spare parts using airlift. The system also includes a network to link forward support locations and CONUS support locations to forward operating locations. MAJCOM components need to work with USTRANSCOM to develop integrated plans to transition peacetime operations smoothly into wartime operations. An expeditionary Air Force cannot afford critical resources to sit backlogged at FSLs and transshipment points.

DoD should explore the option of having a single agency responsible for strategic lift and distribution within theater—an end-to-end military system—as compared with the current system, in which the U.S. Transportation Command and the combatant commander share responsibilities. The difference between a strategic movements system and a tactical movements system is not clear. For instance, is a system that connects CIRFs or supply FSLs located in one AOR to FOLs in another AOR (as happened with CIRF shipments and other supplies in both OEF and OIF) a strategic movements system or a TDS? If it is a TDS, which combatant commander should set up the inter-AOR system, the supporting commander or the supported commander? Perhaps the separation of the TDS and the strategic movements system has outlived its usefulness, given the global war on terrorism and the global positioning of combat support resources to meet commitments across a wide variety of scenarios.

CHAPTER SEVEN

Exploitation of Technology

Technological advances improved warfighting capability in recent operations. In this chapter, we discuss technology used during OIF, as well as how technology may be exploited for future operations.

Findings

Our findings are in the following areas:

- Communications
- Munitions
- Fuels.

Communications

During JTF NA, communications personnel from the Air Force Engineering and Technical Service (AFETS) played a limited, but significant, role. Their role was limited by the fact that a direct UTC did not exist for war planners to use for AFETS and by a limited understanding of the availability of AFETS for deployment. AFETS who did deploy were tasked as a substitute for a military position from the tasked unit.

Before OEF began, a communications AFETS UTC (6KNX4) was developed. This UTC provided a far more robust AFETS tasking capability, and AFETS personnel were among the first in-theater in support of OEF. However, in some cases, obtaining visas impeded rapid deployment into the theater. Deconflicting the specific AFETS skills sets required for theater taskings was also a challenge.

The Engineering and Technical Service (ETS) program office worked with the Secretary of the Air Force and the State Department, and expedited passport and visa processing were granted for AFETS personnel before OIF began. The ETS program office also walked-through high-priority visa requests with the appropriate embassies. Prior to the start of OIF, a communications AFETS cadre was established at Central Command/A-6 to assist with optimizing communication capability in-theater. During the conflict, AFETS communications personnel were managed as a high-demand/low-density asset. ACC/SCC and the Central Command A-6 developed rules of engagement (ROEs) that appointed the cadre as the single focal point for all communications AFETS activities in-theater. The ROEs aided tremendously in deconfliction of taskings and enabled deployed AFETS personnel to be rapidly redistributed within theater to meet changing mission needs.

Since the Air Force had a long time to plan for OIF, communications issues that arose during OEF were addressed before major operations began in support of OIF. In support of OIF, the Air Force bought additional bandwidth for military use prior to the beginning of combat operations. With the additional bandwidth, communications with ISR assets, such as Global Hawk and Predator, became almost real-time, whereas, during OEF, the time between acquiring a target and receiving permission to shoot was often too long to be effective.

The targeting system was much improved during OIF. For example, while flying in the AOR, the Predator was controlled from CONUS (see Figure 7.1). Data gathered by Predator over the combat zone were transferred from the airframe to a satellite, then bounced to a receiver in USAFE, then relayed to Langley AFB, Virginia, where they were analyzed and distributed to several agencies, including the Air Operations Center in the AOR. The identified targets were then matched with shooters. This process required only a few minutes—a great improvement from experiences during OEF.

Figure 7.1
A Sophisticated Communications System Was in Place During OIF

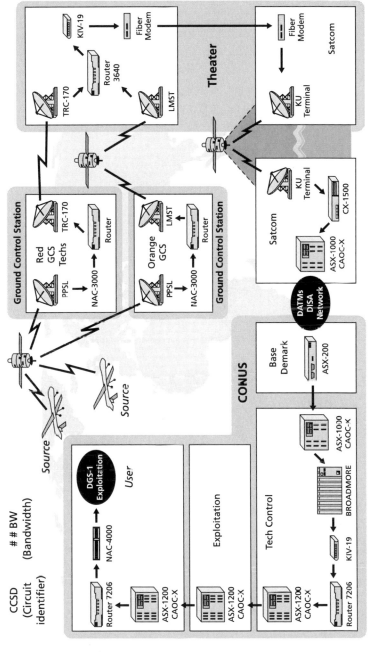
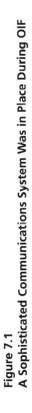

CCSD   # # BW
(Circuit   (Bandwidth)
identifier)

RAND MG193-7.1

Near-real-time communications were applied to most ISR assets. Figure 7.2 illustrates three ISR assets used during OIF.

The Global Hawk was programmed and launched in the AOR. The downloaded data were transmitted back to CONUS and analyzed at Beale AFB, California. The analysis was then sent to Langley AFB, Virginia, where it was distributed.

The Hunter, likewise, was launched and controlled within the AOR. Video was captured and sent to Langley. Unlike with the Global Hawk data, the video was then distributed to Army division headquarters; Hunter video undergoes no analysis.

Air Combat Command was able to provide global broadcast service (GBS) receive suites (RSs) to broadcast Predator video in support of OIF on short notice. In February 2003, ACC received tasking to deploy several RSs. At the time, ACC did not have any RSs to deploy and was not scheduled to receive any until March 2003. In addition, ACC did not have sufficiently trained personnel who could operate and maintain the system. ACC immediately requested

Figure 7.2
State-of-the-Art Near-Real-Time Communications Were Used During OIF

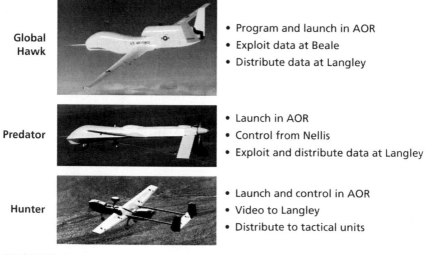

Global Hawk
- Program and launch in AOR
- Exploit data at Beale
- Distribute data at Langley

Predator
- Launch in AOR
- Control from Nellis
- Exploit and distribute data at Langley

Hunter
- Launch and control in AOR
- Video to Langley
- Distribute to tactical units

RAND MG193-7.2

accelerated delivery of RSs, which Air Force Space Command and the Joint Program Office granted. The RSs were shipped to units tasked to deploy. ACC provided Just-In-Time Training, which was initially created to support OEF personnel. Through this effort, ACC was able to deploy tasked units with the capability to support Predator missions.

The communications system in place during OIF was much better than the system in place during OEF. A theater-wide communication plan was developed that included redundant circuits to most locations, with a communication bandwidth increase of nearly 600 percent, and an increase in satellite communication terminals of over 550 percent. The additional bandwidth allowed intelligence data feeds from Global Hawk and Predator to the CONUS; those feeds, in turn, allowed more personnel to stay in the rear instead of deploying forward.

Munitions
Figure 7.3 shows that the use of precision weapons has increased with each recent operation. Because the Global Positioning System (GPS) was so accurate (although not as accurate as laser-guided bombs) and could be used in all weather, the Joint Direct Attack Munition was the favorite. While designed to be used against high-value, fixed targets, JDAMs were heavily used against relatively low-value targets and in close air support missions flown by bombers at relatively high altitudes. The extensive use of precision-guided munitions greatly improved the Air Force's ability to hit targets, in any weather.

The increased use of precision-guided munitions during OEF and OIF translates to less total munitions weight being moved into the AOR. The ability to strike in any weather greatly reduces the time frame for a campaign. Any reduction in time or movement requirement equals a reduction in the forward support necessary in the AOR. As with improved communications, fewer personnel needed to be deployed forward.

Figure 7.3
Precision-Guided Munitions Were Used Increasingly in JTF NA, OEF, and OIF

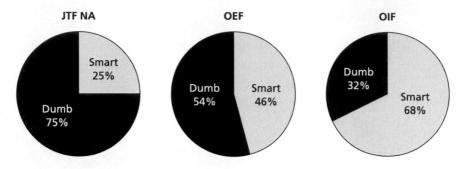

SOURCE: JIF NA and OEF data were abstracted from TPFDDs. OIF data are from
U.S. Air Force, Central Command, 2003.

RAND *MG193-7.3*

Fuels

Fuels is one area in which technology has not been exploited. Air
Force bare base fuels assets use 1960s technology and require extensive spare parts and a large workforce to set them up and operate
them. The current capability requires a combination of pumps, filters, and valves that are not interchangeable and do not employ readily available commercial automation. Better configuration control and
interoperability in maintenance of bare base fuels assets could reduce
both the logistics footprint and personnel footprint.

Many issues that arose during OIF could have been resolved
with simple technology. For example, reports on petroleum, oil, and
lubricants (REPOLs) are not standardized across components. Each
component reports only its own assets. The combatant commander
needs an overall picture. A standardized report for all components
that is submitted to one overall fuels office would fulfill this need.

During both OEF and OIF, deployed Army and Marine personnel were using Air Force fuels mobility support equipment but
were not familiar with the Air Force equipment operation, maintenance requirements, and spares ordering processes.

During OEF and OIF, some local commanders would not allow commercial fuel delivery trucks to have access to the base. Because of terrorist threats, forward locations had to receive fuel at the perimeter of the base. Vulnerable fuel-bladder farms were built at the base perimeter, and the bases were forced to "double handle," or transfer, fuel using limited manpower and flight-line refueling vehicles. This process not only significantly increased force protection but also required additional piping, pumps, fuel bladders, hose line, personnel, paving, and force-protection personnel. A rapidly deployable receipt and transfer hose line system capable of safely transferring fuel from the base perimeter to a secure fuel-storage area would reduce the workload.

Other issues include the size of fuels equipment and hose connections, which currently range from 3 to 6 inches, causing choke points in the fuels-distribution system. Standardization is recommended. There are three types of refueling equipment that require three different sizes and/or types of fuel filters. A single filter that would work for all three assets would decrease the logistics footprint. Likewise, 210K fuel bladders are more effective than 50K bladders, allowing more flexibility in fuels operations while reducing the logistics footprint. Some engines and support equipment have difficulty using JP8 jet fuel. Again, standardization is recommended. Fuels Support Kits (UTC-JFDES), critical to fuels operations at bare base locations, are not sized for ease of movement by military airlift.

Implications

Communications support requirements have changed. No longer are communications limited to just basic bare base systems of local area networks (LANs) and telephone lines. In recent operations, communications personnel were expected to understand the systems and programs LANs connect. Personnel were asked to resolve issues with the Theater Battle Management Core System (TBMCS) and other command and control systems they do not use in garrison on a day-to-day basis. Because of these new requirements, education and training on operating and maintaining command and control systems needs to be developed for communications personnel.

With unmanned aerial vehicles being flown in the AOR by personnel in CONUS, fewer communications and analysis personnel are required to be deployed forward during an operation. For each person not required to be deployed forward, a corresponding number of combat support personnel also do not need to be deployed. Advances in munitions technology can also reduce the deployment footprint.

During the last two operations, OEF and OIF, precision-guided munitions were used increasingly more often. Often fewer smart bombs are required to achieve a target than dumb bombs. Fewer munitions used means less of a deployment footprint, both in terms of the bombs themselves and the associated support equipment and personnel.

Fuels support at bare bases used technology from the 1960s. Extensive spare parts and a large workforce are required to set up and operate the fuels systems, and current equipment is not interchangeable: Each service has different equipment, different training, and different reporting. Better configuration control and interoperability in maintenance of bare base fuels assets in the Air Force and among the services could reduce both the logistics footprints and personnel footprint. Developing a modular, scalable Fuels Support Kit would allow deploying members to hand carry key portions of the kit. Moreover, developing common fuels mobility equipment and training would alleviate on-the-job-training at a deployed site during contingency operations.

By reducing the number of personnel and amount of equipment taken forward to support the warfighter, a corresponding reduction occurs in the amount of services, security forces, etc., reducing the overall footprint.

CHAPTER EIGHT

Resourcing to Meet Contingency, Rotational, and MRC Requirements

Combat support resources are allocated and employed in meeting today's AEF rotational and contingency requirements in ways that are not consistent with the assumptions that are made in the current resource requirements determination processes. In this chapter, we analyze how resource-planning factors and processes may need to be changed to better meet the needs of today's expeditionary air and space forces and current defense programming guidance.

Findings

Our findings are in the following areas:

- Harvest Falcon assets
- Munitions
- Critical Combat Support Personnel.

Harvest Falcon Assets

Table 8.1 illustrates how those planning factors that are used to determine Harvest Falcon requirements (left side of table) differ significantly from how Harvest Falcon assets are employed today (right side of table). As can be seen, the planning factors are based on supporting full-size squadron deployments to a bare base with adequate room to set up Housekeeping, Flight Line, and Industrial Operations sets. JTF NA, OEF, and OIF experiences have shown that numerous Air

Table 8.1
Harvest Falcon Planning Factors Versus Actual Employment Today

Harvest Falcon Resource Planning Factors	Harvest Falcon Current Employment Factors
Bare base deployment; space and latitude to build to economies of scale	Deployment to existing bases to augment infrastructure; must fit in space available
Short, intense wartime involvement; minimal infrastructure to generate sorties	Sustained, indefinite deployments/ employments; additional quality-of-life and force-protection requirements
MRC full-squadron deployments	Less-than-squadron deployments and modular FOL support
High-threat force-protection requirements not included	Significant additional requirements for FOL support modules/items
Support to Air Force units only	Support for other services
Harvest Falcon requirements, FSL, and distribution throughput computed against specific planning scenarios	Harvest Falcon sets have been used to support other AORs routinely—e.g., support of Burgass in OEF, other USAFE sites in JTF NA, and throughput needs to be computed to meet global AEF goals

Force deployments involve deploying in less-than-squadron-size units to coalition-partner military sites. The deploying forces may fall in on existing infrastructure but require additional assets—for example, power distribution units. Also, because of space limits, detached facilities may have to be built in a restricted amount of space. Further, specific components of sets—for example, light sets—are issued to meet specific demands for force protection or other needs. Requirements planning factors also assume that the sets would be used one time to meet major regional contingency (MRC) needs. Today they are being used to sustain long-term permanent rotations.

The last row in Table 8.1 addresses the fact that specific planning scenarios are used to determine requirements, when these assets are really needed to meet AEF support requirements worldwide. It may be that the global goals are more stringent than specific theater needs.

Figure 8.1 shows how the difference in planning assumptions and employment factors created shortages in particular Harvest

Figure 8.1
OIF Harvest Falcon Employment Practices Differ
from Planning Practices

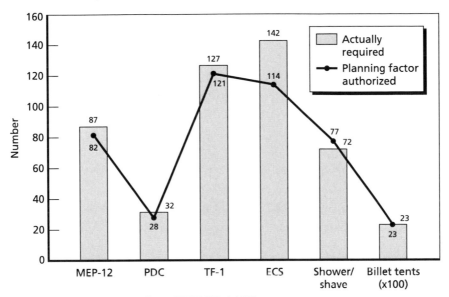

SOURCE: Data are from CENTAF/A-4 LGX.
NOTES: In the fuels arena, there were not enough liquid-oxygen sample cylinders to meet CENTAF sampling requirements. MEP-12 = power generators; PDC = power-distribution centers; TF1 = lighting units; ECS = expandable common-use shelters.
RAND MG103 8.1

Falcon components and in mission-capable and deployable sets during OIF. Specific high-demand components of Harvest Falcon sets are issued to support deployments and are removed from complete sets to meet demands: power-generation (MEP-12 generators, power-distribution centers [PDC], lighting units [TF-1], expandable common-use shelters [ECS]), shower/shave units, and billeting tents. In all but the shower/shave units, demand and operational needs exceeded the planning factor authorizations.

During OIF, some of these shortfalls were alleviated when additional combat support resources were obtained. Additional contract dollars were applied to critical shortages.

The Harvest Falcon example demonstrates two of the differences between programming assumptions and current employment mentioned earlier: (1) that specific commodities, not complete sets, are issued to meet needs and (2) that the commodities are not returned to storage, but tend to remain issued and in use for extended periods. Thus, complete sets are rendered incomplete and not ready for deployment.

Harvest Falcon kits are employed in more than just MRCs. Operations Northern Watch and Southern Watch have also reduced the availability of Harvest Falcon kits.[1] Beddown for Operations Northern Watch and Southern Watch, as well as that for OEF, was using Harvest Falcon kits before OIF even began, which limited the number of kits available for OIF. WRM assets were intended to be used during major regional contingencies. In reality, these Harvest Falcon assets are being used for most operations in the AOR.

Munitions

Table 8.2 shows some of the major disconnects between the munitions resource planning factors used in the process for determining munitions requirements and the actual employment of munitions in current contingency actions, including JTF NA, OEF, and OIF.

The public intolerance for collateral damage has placed a premium on the use of precision-guided munitions (PGMs), such as JDAM, in recent scenarios. The use of precision weapons has increased with each current operation. While designed to be used against high-value fixed targets, precision weapons were heavily used against relatively low-value targets and in close air support missions flown by bombers at relatively high altitudes.

Critical Combat Support Personnel

Disconnects between resource planning assumptions and actual AEF employment factors can create not only resource shortages, requiring

---

[1] Interview with Maj Dennis Long, CENTAF/A-1 LGX, September 2002.

Table 8.2
Munitions Planning Factors Versus Actual Employment Today

Munitions Resource Planning Factors	Munitions Current Employment Practices
Specific scenarios, aircraft types, and target sets are used to compute requirements	Actual scenarios differ from planning scenarios
Precision munitions are determined against specific targets in the scenarios	Precision munitions are the munitions of choice, owing to tight rules of engagement on collateral damage
Computations assume that munitions will be used in specific scenarios and are distributed to specific combatant commanders for anticipated use in specific AORs	Munitions that are distributed to specific AORs are used in other AORs routinely
Munitions FSL throughput and distribution requirements are determined, if completed based on specific scenario considerations—e.g., trucking capacity in Korea	Munitions FSL throughput and distribution capabilities need to be based on global needs

the Air Force to make a concerted effort to reconcile planning and employment factors, but also personnel issues. The Air Force has been deploying forces to Southwest Asia in support of Operation Southern Watch and Operation Northern Watch since the end of Operation Desert Storm. This continuing deployment and the requirement to quickly move forces led to the AEF concept. The management of the AEF, as it has evolved in the Air Force, has become very complex. Prior to OIF, the Air Force AEF model was designed to evenly distribute resources into ten AEF deployment parts, often referred to as "buckets." Each bucket has roughly equal capability. The buckets are paired into five deployment cycles, when a bucket would be eligible for deployment, on a set rotational schedule for three-month rotations. The Combat Air Force aircraft are managed by ACC/DO and the equipment is managed by function. Prior to the implementation of the AEF concept in 2000, Expeditionary Combat Support (ECS) individuals were managed by their respective functional area managers (FAMs) under the Palace Tenure Program. Under the AEF concept, ECS is no longer managed as individual augmentees; it is scheduled as UTCs. The AEF Center nominates, and MAJCOM FAMs verify, sourcing of ECS UTCs to meet combatant commander requirements.

The AEF Center was originally formed to help prepare units for deployment and track problems and lessons learned from these deployments. In 2000, the AEF Center absorbed the ACC ECS Scheduling Integration Team (ACC/XOOS), which was responsible for integrating the scheduling and sourcing of all ECS UTCs under the AEF concept. Continuing operations, including OEF, severely depleted a number of ECS functional areas. As the size and scope of OIF was realized, it quickly became apparent to Air Force planners that the ECS forces that had been apportioned to 1003V were either not available or had recently returned from deployment.

Future temporary-duty pledges had been violated for several career fields even prior to OIF. Figure 8.2 shows the forward-reaching and extended tours for personnel as of December 2002. At the beginning of OIF, the AEF rotations were frozen. With shortages in many high-demand/low-density career fields, personnel in the AOR had to remain in the AOR until the completion of OIF. Part of the problem was that OIF followed so quickly after operations in Afghanistan. Personnel were still deployed to Afghanistan, limiting the pool of personnel available to deploy to Iraq.

The AEF Center was the only Air Force organization ready to provide a clear picture of global ECS capability. It was able to nominate the most ready units to fill requirements. It first looked to UTCs aligned in the current bucket; then, after approval from AF/XO, it looked to the next bucket; and so on until the requirement could be either filled or identified as a service shortfall. The AEF Center found itself in the unusual position of directly nominating into JOPES the sourcing of ECS UTCs for deployment to meet the crisis response.

While this process worked well for personnel UTCs, the same cannot be said for equipment. There is no single source for the global management of equipment. Moreover, different organizations are responsible for different parts of the sourcing process. Because the Air

Figure 8.2
AEF Rotational Cycle Extensions as of December 2002

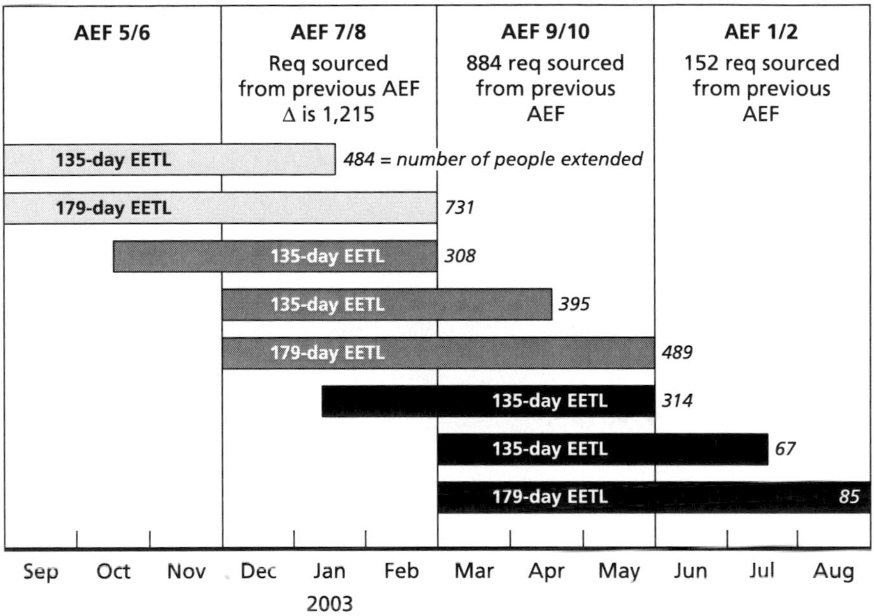

AEF 5/6	AEF 7/8	AEF 9/10	AEF 1/2
	Req sourced from previous AEF Δ is 1,215	884 req sourced from previous AEF	152 req sourced from previous AEF

135-day EETL — *484 = number of people extended*

179-day EETL — *731*

135-day EETL — *308*

135-day EETL — *395*

179-day EETL — *489*

135-day EETL — *314*

135-day EETL — *67*

179-day EETL — *85*

Sep   Oct   Nov   Dec   Jan   Feb   Mar   Apr   May   Jun   Jul   Aug
                              2003

SOURCE: Przybyslawski, 2002, chart 8.
NOTES: Data as of December 10, 2002. EETL = Estimated Extended Tour Length.
RAND MG193-8.2

Force tries to maintain a team concept, they place a burden on ACC/DO and the AEF Center to coordinate taskings. Such coordination was successfully accomplished by using a MAJCOM-approved Target Base Alignment Template to schedule ECS UTCs in the AEFs. This template is based on the Combat Air Force and the Mobility Air Force aircraft schedule; however, it is complicated by the fact that aviation UTCs and ECS UTCs are not scheduled back to back.

Implications

Three recent operations have indicated that the current resource planning factors and methods are not aligned with current resource consumption factors. Combat support resources are stretched thin in meeting current rotational, peacekeeping, and training requirements and may leave little capability for meeting future small-scale contingencies or potential MRCs. We show that small-scale contingencies such as JTF NA, OEF, and OIF may not necessarily require fewer support resources than an MRC. In fact, actual resource employment patterns differ from those used in MRC planning computations; in some cases, small-scale contingencies may actually require as many resources or even more.

One possible answer to the problems of limited resources and planning factors not matching actual employment of resources would be to change the factors and increase the inventory levels of materiel and to add personnel. Perhaps nondeploying or limited-deployment career fields such as the Air Force Personnel Center or instructor slots in the Air Education and Training Command could be used to address shortfalls in stressed career fields. The sourcing issue will only be exacerbated once Homeland Security defense manpower requirements are identified and factored into worst-case unit taskings. Computations could be made to determine requirements as a function of the current combat support posture and policies; however, with many competing needs, the Air Force may not be able to afford this approach. Nonetheless, there are several options and trade-offs between alternative requirements, alternative combat support distribution options, and other support policies that may be able to satisfy operational requirements more effectively than just increasing the size of existing pipelines, assuming that the current ways of providing combat support will continue.

One such option would be to make investments to decrease delivery time. For example, items could be positioned closer to the point of need by perhaps distributing existing resources to more FSLs in differing AORs. Another option to decrease delivery time would be to improve throughput capability of existing FSLs and associated distribution capability. Distribution improvements could be made by increasing working maximum on ground (MOG) at FSL sites or nearby airports, or by improving rail or sea handling capabilities. Additional ships to store and move WRM may improve delivery times to FOLs. Smaller, faster ships packaged with high-demand assets may help to alleviate some initial airlift concerns. An integrated analysis of options is needed.

Planning factors for WRM requirements determination and global WRM distribution capabilities need to be considered jointly. Alternatives to stockpiling munitions and other WRM assets need to be considered in today's uncertain world. One approach may include just-in-time munitions production, beyond having on hand stocks needed to support the initial phases of possible contingencies.

To evaluate combat support options today requires a capabilities-based assessment method. Such a method provides insights into the capabilities that exist to meet a wide variety of scenarios and alternative levels of investments in combat support resources. A capabilities view of resources may be a more appropriate way to consider resource investments in today's world than the current scenario-based method. Using this view, various investments could be stated in terms of what they could support—for example, the ability to support X permanent rotations, a small-scale contingency of Y size (defined by beddown sites), and an MRC of Z size (defined by beddown sites). The Air Force will never know what scenarios it may be expected to support in the future with certainty, but it should have the ability to state what capabilities it can support from a combat support perspective.

Then, systems and organizations need to be developed or refined to enhance expeditionary operations. The AEF Center could function as the sole source to nominate all AEF forces for deployment to include aviation UTCs. While this responsibility may not have been part of the AEF Center's original design, the Center handled it smoothly during OIF and is well suited to do the job. They did not, however, source much of the equipment, the actual aircraft, or the associated flying squadrons. If the functions from ACC/DO were moved to the AEF Center and if the Center was given tools and personnel to manage the equipment issues, it could manage all aspects of the deployment nomination process.

CHAPTER NINE

Conclusions

There are a number of opportunities for improving combat support for the AEF of the future. In this chapter, we summarize those opportunities.

CSC2 processes and C2 organizational alignments have improved since JTF NA and OEF. The implementation of the TO-BE operational architecture has aided in this development. Integrating deliberate planning processes and crisis-planning activities is an area needing more work. While deliberate planning is time-consuming, the process fosters an understanding of the AOR and helps to identify shortfalls. During Crisis Action Planning, there is not always time to do the detailed analysis and coordination required during the deliberate stage. Planners should receive training in deliberate planning so that they are prepared for deliberate and Crisis Action Planning.

Austere FOLs and an immature theater infrastructure emphasized the importance of early planning, knowledge of the theater, and FOL preparation. The Air Force recognizes the need to develop these processes and has taken steps to improve them with a standardized GEOReach ACS site-survey CONOPS. Survey information to develop FOLs was more readily available during OIF because of other ongoing operations in the region. Host-nation support was difficult to negotiate, and resultant deployment timelines varied widely throughout the theater. Without host-nation support and sufficient infrastructure already in place, the 5-day notional AEF goal is largely unobtainable. Perhaps the notional goal should be adjusted to a more attainable goal.

The current AEF force structure of light, lean, and lethal response forces is highly dependent upon FSL capacities and throughput. Austere FOLs and the immature theater infrastructure illustrated the importance of using FSLs efficiently. Improvements have been made in linking FSLs and CSLs to dynamic warfighter needs, but much more can be done in this area.

AEF operational goals depend on assured and reliable end-to-end deployment and distribution capabilities that can be configured quickly to connect the selected sets of FOLs, FSLs, and CSLs in contingency operations. By using CIRFs, the Air Force has traded early strategic lift requirements for a continuous sustainment requirement. CIRFs and other forward support have enabled the combatant commander to deploy more warfighting forces in the place of combat support capabilities. The continued success of CIRFs relies on dependable resupply.

Since the Air Force may be the predominant user of the theater distribution system in early phases of future campaigns, the Air Force may be delegated the TDS responsibility. Even if another service is delegated TDS responsibilities, the Air Force should play an active role in determining TDS capacities and capabilities. The Air Force has made advances in the use of centralized maintenance, expanding its dependence on FSLs to provide support; yet, it finds itself poorly prepared to estimate lift requirements.

Current joint doctrine places the responsibility for the development of the strategic movements system on USTRANSCOM; intratheater lift is the responsibility of the combatant commander for the AOR.[1] Thus, current doctrine splits the responsibility for developing the end-to-end deployment and resupply system among multiple par-

---

[1] The Secretary of Defense named USTRANSCOM the Department of Defense Distribution Process Owner on September 16, 2003. A pilot program is currently under way to field a Deployment/Distribution Operations Center (DDOC) to maintain visibility of movements, direct intratheater movement, synchronize strategic/operational lift, and track movement from origin to supply support activity.

ties. Having combat support facilities located in one AOR supporting a combatant commander in another AOR—for instance, moving WRM or repaired spares from the European Command AOR to the Central Command AOR—confuses TDS and strategic movements. When the strategic system and the theater distribution system came together at transshipment points, there were significant backlogs and system disconnects.

This joint doctrine may be inappropriate for expeditionary forces, which rely on fast deployment, immediate employment, and responsive resupply of lean forward-deployed forces. The Air Force's reliance on lean deployments and responsive resupply of deployed units places great importance on the rapid development of contingency end-to-end deployment and distribution capabilities.

Significant improvements in communications were achieved during OIF. Near-real-time raw intelligence data were received in CONUS, and the data were exploited and then redistributed to numerous agencies; at the same time, personnel were identifying emerging targets and coordinating attacks, all from inside CONUS. These communications advances reduced the number of expeditionary combat support personnel and equipment that had to be deployed, and more resources were kept in the rear. Other opportunities to exploit technology may present themselves.

Finally, the planning factors and assumptions that are used to determine resource requirements differ significantly from those that are encountered in current rotational and contingency operations. In many cases, the current resource employment factors are more demanding than the assumptions used to fund resources. This imbalance creates the resource shortages that appear in contingency operations. Shortages in combat support assets—particularly those in high-demand/low-density areas, such as combat communications, civil engineers, and force protection—stressed the AEF construct.

In addition, the current AEF employment practices differ significantly from planning factors used in the POM process to provide for combat support resources. The current AEF scheduling rules are routinely violated in stressed combat support areas. Current AEF scheduling rules may be an effective and efficient way to schedule and deploy aircraft and aircraft support units, but the current rules may not be the best for scheduling combat support. Scheduling rules should afford ACS personnel opportunities to train to wartime skills and ACS leaders to train to properly employing those personnel. Specifically, balances should be struck between disrupting home-station support and meeting deployment commitments.

Below is a list of the recommendations derived from the work on this study. The recommendations are categorized into two sections: those that can be achieved with near-term actions and those that can be achieved with long-term actions by the joint community, Congress, and the Department of State. These recommendations are suggested methods to improve Agile Combat Support for the AEF.

Near-Term Recommendations

Combat Support Execution Planning and Control (CSC2)

- Integrate deliberate and crisis planning activities.

Forward Operating Locations and Site Preparation

- Standardize site-survey procedures and processes within the Air Force, with U.S. allies, and with other services.

Forward Support Location/CONUS Support Location Preparation for Meeting Uncertain FOL Requirements

- Continue improvements in linking FSLs and CSLs to dynamic warfighter needs.

Reliable Transportation to Meet FOL Needs

- Ensure dependable resupply to centralized intermediate repair facilities.

- Identify lift requirements, including airlift, sealift, and movement by land, for theater distribution system.
- Provide additional training and enhance personnel development policies for the Air Force to meet future theater distribution responsibilities, such as in the exercise EAGLE FLAG.

Exploitation of Technology

- Review contingency combat support functions that could be done in the rear (CONUS) because of advances in communications technology processes—for example, sustainment planning and execution—that offer the possibility of reducing the forward-deployed footprint.

Resourcing to Meet Contingency, Rotational, and MRC Requirements

- Evaluate existing combat support scheduling rules with respect to impacts on home-station and deployed combat support performance.

Long-Term Recommendations

Combat Support Execution Planning and Control (CSC2)

- Consider requirements of the joint and unified commands and identify how to meet those requirements while remaining responsive and adaptive.

Forward Operating Locations and Site Preparation

- Focus attention on political agreements and engagement policies.

Forward Support Location/CONUS Support Location Preparation for Meeting Uncertain FOL Requirements

- Further develop the existing global network of FSLs and CSLs.

Reliable Transportation to Meet FOL Needs

- Review joint doctrine on the transportation system.
  —Consider having USTRANSCOM develop end-to-end distribution channel capabilities.
  —Consider ways to improve TDS performance, including examining pricing mechanisms, and instituting better in-transit visibility and demand-forecasting mechanisms.

Resourcing to Meet Contingency, Rotational, and MRC Requirements

- Reevaluate current processes and policies for AEF assignments and the current POM assumptions with respect to combat support resources.

Combat Support Execution Planning and Control (CSC2) TO-BE Operational Architecture

In the CSC2 TO-BE operational architecture (Leftwich et al., 2002), a CSC2 nodal[1] template is established. Table A.1 presents the CSC2 nodal responsibilities and processes outlined in the TO-BE operational architecture.

Table A.1
TO-BE CSC2 Nodes and Responsibilities

Combat Support C2 Nodes	Roles/Responsibilities
Joint Staff	
Logistics Readiness Center	Supply/demand arbitration across combatant commanders
Combatant Commander	
Combatant Commander Logistics Readiness Center	Combatant commander logistics guidance and Course of Action analysis
Joint Movement Center	Combatant commander transportation supply/demand arbitration
Joint Petroleum Office (JPO)	Combatant commander POL supply/demand arbitration
Joint Facilities Utilization Board	Combatant commander facilities/real estate supply/demand arbitration
Joint Materiel Priorities and Allocation Board	Combatant commander materiel supply/demand arbitration

---

[1] A *CSC2 node* is a point of intersection, within a larger infrastructure, where the integration of processes and information occurs.

Table A.1—Continued

Combat Support C2 Nodes	Roles/Responsibilities
JTF	
JTF J-4 and Logistics Readiness Center`	JTF logistics guidance Supply/demand arbitration within JTF, among service components
JFACC	
Joint Air Operations Center Combat Support Reps	JAOP/MAAP/ATO production support
JFACC Staff Logisticians	JFACC logistics guidance
Air Force	
Air Force Contingency Support Center (CSC)[a]	Monitor operations Represent Air Force combat support interest to Joint Staff Conduct/review assessments of integrated weapons systems and base operating support Arbitrate critical resource supply/demand shortages across AFFORS
AFFOR	
Air Operations Center (AOC) Combat Support Element	JAOP/MAAP/ATO production support
AFFOR A-4 Staff (forward)	Site surveys/beddown planning Liaison with AOC combat support element
AFFOR A-4 Staff (rear) at an Operations Support Center (OSC),[b] which supports AFFOR A-4 Staff (forward)	Mission/sortie capability assessments Beddown/infrastructure assessment ASETF force structure support requirements Supply/demand arbitration within ASETF among AEFs/bases Theater distribution requirements planning Force analysis of speed of delivery Liaison with Air Mobility Division in AOC Liaison with theater USTRANSCOM node
Deployed Units	
Wing Operations Center (WOC)	Disseminate unit tasking Report unit status
Combat Support Center	Monitor and report performance and inventory status
Supporting Commands (Force and Sustainment Providers)	
Logistics Readiness Center/CSC	Monitor unit deployments Allocate resources to resolve deploying unit shortfalls

Table A.1—Continued

Combat Support C2 Nodes	Roles/Responsibilities
Deploying Units	
Wing Operations Center (WOC)	Report unit status Disseminate unit tasking
Deployment Control Center (DCC)	Plan and execute wing deployment Report status of deployment
Commodity Control Points (CCPs)[c]	
Munitions, Spares, POL, Bare Base Equipment, Rations, Medical Materiel, etc.	Monitor resource levels Perform depot/contractor capability assessments Work with the CSC to allocate resources in accordance with theater and global priorities
Sources of Supply (Depots, Commercial Suppliers, etc.)	
Command Centers	Monitor production performance and report capacity

[a]Some of these functions, which will be performed by the CSC, were referred to as Global Integration Center (GIC) functions in Leftwich et al. (2002). The Air Force will not use the GIC name in implementation efforts; rather, it will associate GIC functions with the CSC.

[b]The functions performed by the AFFOR A-4 forward and rear need not be the same for all theaters or regions. The idea is to codify the responsibilities by COMAFFOR in each region before contingencies begin. OSC A-4 will have virtual Regional Supply Squadron representation at the OSC. Many of the spares-related command and control functions would be conducted at the RSS with OSC A-4 input and coordination. The same is true for ammunition control points.

[c]The CCP was referred to as a Virtual Inventory Control Point (VICP) in Leftwich et al. (2002) and in several articles associated with the Spares Campaign and Depot Reengineering and Transformation. CCP will replace the VICP name.

Bibliography

AFCAP Update, "Support to USAFE and Operation Sustain Hope," informational brochure published by Readiness Management Support L.C., n.d. Available at www.rms-lc.com/updates/ShiningHope/

Air Force Materiel Command, *Air Force Materiel Command Warfighter Sustainment Division Concept of Operations,* March 2003.

Amouzegar, Mahyar A., Lionel A. Galway, and Amanda Geller, *Supporting Expeditionary Aerospace Forces: Alternatives for Jet Engine Intermediate Maintenance,* Santa Monica, Calif.: RAND Corporation, MR-1431-AF, 2002.

Barthold, Col Bruce R., "Major Issues from the AFCESA/AFIT Sponsored Operation Enduring Freedom RED HORSE and Prime BEEF Lessons Learned Conference, 13–15 Nov 02," presented at conference, Dayton Ohio, December 2002.

Boyd, John R., "A Discourse on Winning and Losing," Maxwell AFB, Ala.: Air University Library, Document No. M-U43947, unpublished collection of briefing slides, August 1987.

Burns, Brigadier General Patrick, "U.S. Air Force Expeditionary Combat Engineering in Operation Enduring Freedom," briefing, 2003.

DynCorp, "Operation Enduring Freedom Oct 01–Mar 02," briefing to DynCorp staff, Shaw AFB, South Carolina, n.d.

Feinberg, Amatzia, H. L. Shulman, L. W. Miller, and Robert S. Tripp, *Supporting Expeditionary Aerospace Forces: Expanded Analysis of LANTIRN Options,* Santa Monica, Calif.: RAND Corporation, MR-1225-AF, 2001.

Galway, Lionel, Mahyar A. Amouzegar, R. J. Hillestad, and Don Snyder, *Reconfiguring Footprint to Speed Expeditionary Aerospace Forces Deployment*, Santa Monica, Calif.: RAND Corporation, MR-1625-AF, 2002.

Galway, Lionel, Robert S. Tripp, Timothy L. Ramey, and John G. Drew, *Supporting Expeditionary Aerospace Forces: New Agile Combat Support Postures*, Santa Monica, Calif.: RAND, MR-1075-AF, 2000.

Headquarters U.S. Air Force (HQ USAF/IL), External Slides, daily update to CSAF, September 19 and 20, 2001a. Posted at www.afxo. pentagon.smil.mil/

HQ USAF, Operations Group briefing, October 30, 2001b.

HQ USAF, *USAFE/SA Air War Over Serbia Report*, April 2000.

Hillestad, Richard, Robert Kerchner, Hyman Shulman, and Louis Miller, "The Closed Loop System for Planning and Executing Weapon System Support," Santa Monica, Calif.: RAND Corporation, briefing to RAND staff, January 13, 2003.

Joint Chiefs of Staff, *Joint Tactics, Techniques, and Procedures for Movement Control*, Washington, D.C.: Joint Publication 4-01.3, April 9, 2002.

Jumper, Gen John P., "The Limitations of Doctrine," briefing given to Air Force Doctrine Center, Maxwell AFB, Ala., n.d.

Krugman, Paul, "Thanks for the M.R.E.'s," *The New York Times*, August 12, 2003.

Lambeth, Benjamin S., *NATO's Air War for Kosovo: A Strategic and Operational Assessment*, Santa Monica, Calif.: RAND Corporation, MR-1365-AF, 2001.

Leftwich, James A., Robert S. Tripp, Amanda Geller, Patrick H. Mills, Tom LaTourrette, Charles Robert Roll, Cauley Von Hoffman, and David Johansen, *Supporting Expeditionary Aerospace Forces: An Operational Architecture for Combat Support Execution Planning and Control*, Santa Monica, Calif.: RAND Corporation, MR-1536-AF, 2002.

Peltz, Eric, H. L. Shulman, Robert S. Tripp, Timothy L. Ramey, Randy King, and John G. Drew, *Supporting Expeditionary Aerospace Forces: An Analysis of F-15 Avionics Options*, Santa Monica, Calif.: RAND Corporation, MR-1174-AF, 2000.

Przybyslawski, Brig Gen Tony, "AEF Weekly Metrics." briefing dated December 10, 2002.

Pyles, Ray, and Robert S. Tripp, *Measuring and Managing Readiness: The Concept and Design of the Combat Support Capability Management System,* Santa Monica, Calif.: RAND Corporation, N-1840-AF, 1982.

Rosenthal, Maj Robert, "Air Force Operations Group Noble Eagle/ Enduring Freedom Operations Update," briefing, January 15, 2002.

Tripp, Robert S., Lionel A. Galway, Paul S. Killingsworth, Eric Peltz, Timothy L. Ramey, and John G. Drew, *Supporting Expeditionary Aerospace Forces: An Integrated Strategic Agile Combat Support Planning Framework,* Santa Monica, Calif.: RAND Corporation, MR-1056-AF, 1999.

Tripp, Robert S., Lionel A. Galway, Timothy L. Ramey, Mahyar A. Amouzegar, and Eric Peltz, *Supporting Expeditionary Aerospace Forces: A Concept for Evolving to the Agile Combat Support/Mobility System of the Future,* Santa Monica, Calif.: RAND Corporation, MR-1179-AF, 2000.

Tripp, Robert S., Kristin F. Lynch, John G. Drew, and Edward W. Chan, *Supporting Air and Space Expeditionary Forces: Lessons from Operation Enduring Freedom,* Santa Monica, Calif.: RAND Corporation, MR-1819-AF, 2004.

U.S. Air Force, *Combat Support,* Washington, D.C., Air Force Doctrine Document 2-4, October 2000.

U.S. Air Force, *Command and Control,* Washington, D.C., Air Force Doctrine Document 2-8 (Draft), February 16, 2001.

U.S. Air Force, *Organization and Employment of Aerospace Power,* Washington, D.C., Air Force Doctrine Document 2, September 28, 1998.

U.S. Air Force, Central Command, *Operation Iraqi Freedom—By the Numbers,* report by CENTAF Assessment and Analysis Division, Prince Sultan Air Base, Saudi Arabia, April 30, 2003.

Wolff, Robert D., *Contingency Contracting in Support of Air Expeditionary Forces,* report to RAND under Contract F49642-96-C-001, December 29, 2000.

Wood, David, "Some of Army's Civilian Contractors Are No-Shows in Iraq," Newhouse News Service, August 1, 2003.